A HARVEST

First published in 2005 by Boglark Press
This edition published in 2007 by
Liberties Press
Guinness Enterprise Centre | Taylor's Lane | Dublin 8 | Ireland
www.LibertiesPress.com
General and sales enquiries: +353 (1) 415 1224 | peter@libertiespress.com
Editorial: +353 (1) 402 0805 | sean@libertiespress.com

Distributed in the United States by
Dufour Editions
PO Box 7 | Chester Springs | Pennsylvania | 19425

and in Australia by
James Bennett Pty Limited | InBooks
3 Narabang Way | Belrose NSW 2085

Liberties Press is a member of Clé
the Irish Book Publishers' Association.

Trade enquiries to CMD Distribution
55A Spruce Avenue | Stillorgan Industrial Park | Blackrock | County Dublin
Tel: +353 (1) 294 2560 | Fax: +353 (1) 294 2564

ISBN: 978–1–905483–30–3

2 4 6 8 10 9 7 5 3 1

A CIP record for this title is available from the British Library.

Printed by Athenaeum Press | Dukesway | Team Valley | Gateshead | UK

A Harvest

New, Rare and Uncollected Essays

Con Houlihan

To the memory of Joe Dooley,
Footballer, Printer,
Gentleman, Gentle Man

CONTENTS

AUTHOR'S NOTE

Some of the pieces in this book have already appeared in collected form. They are included to illustrate other pieces. You will find repetition and even contradiction. Bear with me.

ACKNOWLEDGEMENTS

I wish to thank the Evening Press, Magill and the Sunday World for allowing me to use pieces from their publications, Joe Dooley and his staff in Harcourt Printing for their kindness and consideration, and Harriet Duffin for helping with the proof-reading.

FOREWORD

When my son was eleven he asked me to read him The Lord of the Rings. He loved the part where the wandering hobbits encounter the windswept figure of Strider in a remote inn, a mysterious, nomadic figure who underneath his mud-caked travelling cloak is revealed to be Aragorn, the uncrowned true king. That same week – my son sharing my twin interests of literature and sport – we went to see Shelbourne play Bohemians in Tolka Park. The atmosphere was intense, fuelled by local rivalry. Stewards patrolled the pathway behind the goal, preventing anyone from leaning on the rail there to observe the game at close hand. As we were ushered up into the stand, my son's attention was taken not by the frantic action on the pitch, but by a quiet figure, with his hand half-covering his face, who leaned on that forbidden rail behind the goal to intently observe the action. Anybody else who dared pause there was bustled away, but no steward had the audacity to approach this man, to whom ordinary rules did not apply. 'Who is that, Dad?' my son asked. 'That,' I replied, 'is true nobility, that is Con Houlihan, that is Aragorn.'

The metaphor was obscure, but my son instinctively understood. Behind the false glamour of instant celebrity, there is a different type of fame – a true and lasting respect. In an era when the word 'fame' is debased, only a small handful of people are legendary. Other fame can be rewarded with baubles, but match stewards recognising an aura as too special to be disturbed is as good an example of genuine respect as I can think of.

The positioning of Houlihan seemed appropriate at that match. Not up in the press box or among the crowd with their chants, but utterly engaged and utterly apart, standing – as Cavafy said a poet should stand – at a peculiar angle to the universe. Although Houlihan is best known as a sports journalist, he has always had the gift to contextualise sport within the wider tapestry of the human condition, on which he is an expert.

Great sport – like great poetry and great art – is about moments when time stands still. Such occasions are not a matter of life and death, but are more important because they transcend life and death, making us forget everything in the magic of being caught up and enraptured in an extraordinary passion.

Con Houlihan has always possessed that passion and an uncynical enthusiasm for painting and poetry and novels, a way of seeing them afresh and of making them come alive for a reader. I read many of the pieces gathered in A Harvest in the formative years of my life. I still recall my excitement at opening the Evening Press; there, amid the contemporary news that would quickly grow stale, was one man's commentary on what Pound called 'the news that stays news' – great literature and great art. I remember reading about Francis Ledwidge at a time when few people discussed him, and Houlihan's encounter with the paintings of Ferdinand Hodler on a visit to an art gallery in Berne after having watched the Irish and Swiss soccer teams clash 'with all the force of two pillows colliding'.

Houlihan's range is wide – from Knocknagow to Van Gogh and Cézanne to Hopkins – but he makes the artists and writers he admires come alive. He makes you feel his enthusiasm and makes you want to read and experience their work not as rarefied, obscure mysteries but as everyday wonders to be discussed in a pub as just one of the strands that make life both ordinary and extraordinary.

To find the best of those fugitive occasional pieces harvested together here is like stumbling into a gathering of old friends in the company of a master storyteller with a wry eye and a sharp wit – a writer not blinded by fads or fashions but with an ability to see everything with both an innocence and a knowing eye, and to cut to the essence of it in a way that any reader can understand.

Dermot Bolger, August 2007

INTRODUCTION

I can still see the little red Ford Anglia van screeching to a halt outside Mike Lyon's shop in Ballyhaunis, County Mayo, on Monday, Wednesday and Friday evenings in the year of Our Lord, 1977.

About half a dozen of the town's citizens, whom I would loosely describe as kindred spirited, would be gathered outside the shop in Abbey Street discussing the affairs of the day, the nation and the world. The banter would be lively, laced with wicked wit and a searching critical analysis of anyone unfortunate enough to pass by the gathering.

Occasionally, one of our group would walk out into the middle of Abbey Street and strain to look for a sign that the Evening Press delivery van might be approaching from under the railway bridge. I can still see the tall figure of Tom Gilmore, the painter and decorator with a razor wit, shaking his head in the middle of Abbey Street; an acknowledgement that there was still no sign of the red express.

When the red Anglia finally arrived, in a blur and with a screeching of brakes, the young man at the wheel flung a bundle of evening papers onto the sidewalk and roared off into the evening towards Claremorris, his next port of call.

It was usually Sean Ruane, for years the loyal farmhand at the local convent and a fox hunter supreme, who produced the penknife to free the bound package of its treasure. We would grab for the papers like children at a sweet counter and, as one, go straight for the back-page sports column, where Con Houlihan reigned supreme.

The individual members of our little group would only skim the column before carefully folding the paper and branching off homeward in different directions. Con's column was sacred. It could only be read in the comfort of one's home with a full pot of tea on the brew.

When we met again the following day at lunchtime outside Joe Regan's Corner Bar in the middle of our town, the Con

1

Houlihan column was the full focus of our discussion. I remember those afternoons and evenings as times of beautiful innocence – a period of my life when I had just reconnected with my hometown after a four-year sojourn in the hills of Tennessee.

It was my uncle, Michael Mannion, a sports fanatic and lover of fine literature, who introduced me to Con Houlihan's Evening Press sports column and the full-page 'Tributaries' feature that appeared in the paper every second Tuesday.

The uncle shortened many a day in the bog holding court on Con's wisdom and his rare writing talent. 'He has a way with words, a way with words,' the uncle would repeat as he flung the sodden sods of turf at me off his slean. 'He brings you the whole game and a whole lot more. His column is better then listening to the game on the wireless.'

On many a winter evening during the year of 1977 I found my own delight in reading Con Houlihan's 'Tributaries' – that once-a-fortnight focus by Con on the life and work of some of the great artists and writers.

I was just out of university in America, where literature and creative writing in particular had been my passions. I devoured Con's delightful and hugely insightful pieces on my favourite American writers, Walt Whitman, Thomas Wolfe, Sherwood Anderson, James Agee, Ernest Heming-way and F. Scott Fitzgerald. Con's wonderfully crafted essays on artists like Paul Cézanne, Vincent Van Gogh, L. S. Lowry and indeed Derek Hill greatly broadened my own cultural horizons.

During those dark winter nights in Ballyhaunis, Con Houlihan also put me back in touch with John Millington Synge, Sean O'Casey and Charles Kickham, whom I had only read in my youth. His essays brought me to reading again the poetry of Patrick Kavanagh, Francis Ledwidge and Dylan Thomas. He opened a window too, for me, into the world and literature of Thomas Hardy and D. H. Lawrence – who are now firm favourites. Years later, I could fully understand why Con had once received a letter from a grateful Dubliner who wrote: 'Con, you gave me my third-level education.'

When I came back to live in Dublin in the late seventies, I felt that although I had never met him, I already knew Con Houlihan well. I'd see the gentle giant from Castleisland walking along

Burgh Quay on afternoons shortly after the Evening Press had hit the street, and I would often watch in awe from the wings as he held court at the bar in Mulligan's of Poolbeg Street.

Then one wet day in Limerick in 1979, I watched John Treacy win the world cross-country title on a mud-soaked Limerick racecourse; that was my first real meeting with Con – the beginning of a beautiful and enduring friendship.

The following day Con wrote about the race on the back page of the Evening Press. The piece was headlined 'THE MAGIC FOX THAT GOT AWAY'. I can still recall one passage from the piece in particular:

On and on he went until in the mist and rain he was away out in front like Tied Cottage in the first three-quarters of the Gold Cup.

And like Tied Cottage he came down – but it was only a slip at a splashy bend and in a few seconds only his muddy knees reminded one of it. In the last mile, as the powerful Pole, Mallinowski, began to make up ground, he seemed like a leading dog in a very scattered pack.

But the magic fox never looked like being caught – his biggest danger was the tumult of small boys that went out like tugboats to meet him.

As in so many of his columns over the years, Con captured completely the true essence of that great day in Limerick long ago.

What other writer would lead off a piece on Vincent Van Gogh with:

It may seem ridiculous to compare Lestor Piggott and Vincent Van Gogh: each, however, was consumed by a passion that led to sacrifices so enormous that 'normal' people can hardly comprehend.

That opening paragraph immediately draws you into a wonderfully insightful essay on Van Gogh, the unquiet spirit.

Con Houlihan has always had a wry sense of humour, but you suspect too from reading him that, although his experince of sadness and loneliness has not been different in kind from that of other men, it has been sharper in intensity. It may well be a key to his ability to give us such insight into the often-complex world of great writers.

I particularly love a piece he wrote about the last day he worked on the bog in Casleisland:

> I knew it was the last day: I was about to depart for a different world. It was also the last day that I worked with my father.
>
> At about six o'clock we raked the embers of the fire together and quenched them with what water we had left over and with what tea remained in the kettle. I was pierced with an infinite sadness.

I have enjoyed many happy hours in Con's company over the years and he has been hugely supportive of my attempts to keep Irish Runner magazine afloat for the past couple of decades. We have spent a few Christmas days together in Con's house in Portobello – the place I first heard Con sing what I still believe was the best-ever rendition of 'The Fields of Athenry', long before the song went 'public'.

Con Houihan left his beloved Castleisland many years ago and for several decades his back-page column was the flagship of the Evening Press. It broke Con's heart when the Press closed. Burgh Quay was his village and his heartland. He had succeeded his great hero Joe Sherwood as 'King of the Back Page'. Writing a tribute piece recently in Magill magazine to that same Joe Sherwood, Con wrote in appreciation of Joe's work: 'Writing with honesty and insight, you will be respected.'

When I think back to those days in the Ballyhaunis of 1977 and my little group of friends, some now departed, I remember how, in writing with that same honesty and insight, Con earned our respect – as well as that of thousands of readers the length and breadth of the country. The collected pieces between these covers are rare treasures.

Now read on . . .

Frank Greally

A Harvest

Winesburg, Ohio - 'A town with a roof over it'

Biographers are a bold breed: they need to be - when you set out to explore someone's life, you are venturing into territory where some of the signposts are meant to mislead; blessed is he who hasn't experienced humiliation he wished to remain hidden.

Writers are probably the most difficult subjects of all: they tend to dramatise themselves and do not consistently subscribe to the belief that truth is stranger than fiction.

Sherwood Anderson is a notorious example: the story of his epiphany on the road to Damascus is the stuff of high romance - but it isn't true.

In his own telling, he walked out one morning from a successful business and a wife and family and the Ohio town where he was a pillar of society and set out for Chicago to live in poverty and in loneliness and try to write.

"I had been walking too long in a dry river-bed and I wanted to wet my feet again" - so he said.

And many believed him - or perhaps chose to believe: we need the romantic almost as much as we need salt.

Truth may be stranger than fiction but in the case of Anderson's turning point it was rather prosaic.

One morning in his thirty-seventh year he left his wife a rather cryptic note: "There is a bridge over a river with cross-ties before it; when I come to that I'll be all right. I'll write all day in the sun and the wind will blow through my hair."

Then he went to his office where in the middle of dictating a letter he walked out.

Four days later he was found in Cleveland, still in his business suit but unkempt and unshaven and liberally spattered with mud.

He was suffering from what is now called a fugue condition; he wished to get away from 'the American disease - the obsession with comfort and money.'

When he recovered from what plain people would call his nervous breakdown, Anderson went back to his old trade as a writer of advertising copy, settled in Chicago - and kept very much in touch with his family.

The image of the writer living in a single room and existing on bread and onions is his own creation.

It doesn't matter; what does is that he realised his dream, so much so that he is deemed the father of American writing.

The honour is rather exaggerated; Anderson didn't come out of nothing - he recalls the first half of a comment made about Paul Cézanne - "A man whom it took generations to make and whom it will take generations to understand."

Mark Twain was an obvious influence; so was Emerson, if only for his passionate belief that American writers shouldn't be in awe of European literature.

Emerson stressed that much of American writing was little more than words cleverly strung together; it wouldn't pass 'the secret test', a term that is more impressive than explainable.

Anderson was obsessed with the need to articulate the poetry of the commonplace; he would have agreed with Thomas Hardy - "At the graveside of even the humblest man you see his life as dramatic."

The remake of time is a great distiller - even the simplest photograph grows fascinating with age.

Hardy's novels are all set in the world that his parents had known in their youth, Wessex, before the coming of the railway effected profound social and economic change.

It was a simple world - or at least it seemed to be; Anderson's work too is elegiac - he wrote of America 'when every small town had a roof over it.'

He longed for the frugal society in which he had grown up - and in *Winesburg, Ohio* he expressed it in a string of loosely interlinked stories, uneven in quality but powerful in the total effect.

The book hasn't heroes or heroines; the young reporter who figures in several of the stories is a fumbler whose innocence sometimes leads him into humiliation.

The book also lacks overt drama: nothing much happens except that people experience such minor ills as loneliness and frustration and misunderstanding and depression and melancholy and unrequited love.

Winesburg, Ohio is written in language as homely as its themes; writing can never be the same as speech - if only because no speaker uses semicolons - but Anderson's comes about as close as can be.

Not for nothing had he been a successful copy-writer; his simple language is powerfully seductive - you forget that you are reading; you are back in a small town in The Mid-West of America in the second half of the nineteenth century.

Men talk about baseball and horses and women; women talk about religion and jam-making and men.

Boys and girls venture out into the country in wagon-loads to pick blackberries; the seasons come and go; people grow old nourishing unrealistic dreams.

And suffusing all is a tremendous sense of place and time; this is the world in which, as Thomas Wolfe said, the most American sound was the whistle of a distant train.

It is also the world that aroused hopes and ambitions that couldn't easily be fulfilled.

In the most celebrated story we meet a farm worker who in youth had harboured great dreams but who is spancelled by an early marriage.

A young comrade approaches him for advice about his own relationship with a girl.

The older man is particularly conscious of his youthful vision - because it is Autumn and the country around Winesburg is washed with lovely colours.

And in his sadness he dwells in a no-man's-land between the practical and the romantic; the title of the story, *The Untold Lie*, is a brilliant summary.

That little tale is often cited as an example of an epiphany, a sudden revelation.

Anderson was obsessed with epiphanies - and the finding of the real self.

"Man", said Rousseau, "is born free and is everywhere in chains". Anderson would say that the real chains are other people's ideas and opinions.

Van Gogh nourished a kindred intuition; he wrote about the necessity to peel the layers of skin from his eyes.

Anderson was convinced that America had gone the wrong way.

"We have taken as lovely a land as ever lay out of doors and put the old stamp of ourselves on it for keeps."

Thomas Wolfe put a similar sentiment in a rather mystical way: "We are searching for a door, a key; we are lost but we will be found!"

America may have been the land of the bold and the free - but most of the bold and the free are now in reservations.

Sherwood Anderson cared little about style or form - he was about the least 'literary' writer of all time.

He wasn't a particularly nice man; indeed he was a monstrous egoist.

He had, however, enormous moral and mental courage and was totally honest.

He left behind a shoal of books but only two stand up - *Winesburg, Ohio* and the collected stories about horse-racing.

In all his other books you will find marvellous fragments - but too often they are embedded in woefully indisciplined writing.

Cézanne, Lonely Genius

Herbert Read, despite being a native of Yorkshire, became greatly respected as an art critic and a philosopher at large. Nevertheless, he coined the occasional memorable one-liner such as: "The past of every reasonable man is strewn with dead enthusiasms."

It is indeed — and with live enthusiasms too. A long time ago when I was just starting out in secondary school, I was infected with an enthusiasm that seems fated to last the rest of my life.

While reading through a periodical called *The Saturday Book*, I encountered a picture that smote me with an extraordinary sense of wonder.

It was a drawing of a mountain in black and white; from it seemed to emanate a vibrant sense of life — I was entranced. It was the work of a man whose name meant nothing to me but that name became a mantra to brighten many a grey day.

Many years later when some of his sketches were on exhibition in The Municipal Gallery in Dublin, two of us made the pilgrimage from Kerry in a van that had seen better days. Greater love hath no men.

Paul Cézanne was born in Provence in 1839. His father was a rare being, a liberal banker. He understood his son's decision to quit his law studies and embark on a career as an artist and he saw to it that he would never lack for money.

Thus Cézanne doesn't fit the popular image of the artist living in an attic and surviving on bread and onions and cheap red wine.

Cézanne was fortunate in another context: in his adolescence he came to know a youth who was as passionate as himself in his ambition to rise above the common world.

Paul Cézanne and Emile Zola walked together and swam together and fished together and dreamed of living in Paris and devoting their lives to fructifying their genius.

A letter from Zola to Cézanne captures the quintessence of youthful dreams.

"For ten years, we have discussed art and literature. We have lived together and often daybreak found us still talking, searching the past and questioning the future.

"We were full of tremendous ideas. We examined and rejected all systems and agreed that apart from the powerful life of the individual, all else is trivial and worthless." That letter was written in 1866. Zola was 26; Cézanne was 27.

Zola was living in Paris; Cézanne was still in Provence. Zola's original ambition was to become a political journalist but he could find no outlet for his ideas — indeed he was scorned.

He turned to drama with a similar lack of success. Cézanne was suffering his own frustrations. When that letter was written, the two friends faced a bleak future. The dreams of young manhood were in dust.

Eventually Zola gained recognition as a novelist; by the time he was 35 he was established. Cézanne was still struggling in obscurity. That letter was to have a bitter sequel: Zola published a novel, 'L'Oeuvre', which depicted Cézanne as a heroic failure.

It is impossible to account for this act of betrayal. It couldn't have originated in jealousy or begrudgery. It hurt Cézanne all the more deeply because he knew that it was an honest appraisal. It would be easy to say that time took its revenge on Zola and that his fame was waning as Cézanne's was waxing but it wouldn't be true.

Cézanne's fame came after his death; some of Zola's novels, especially 'Germinal' and 'L'Assommoir', read as well today as when they were first published.

How could Zola be so wrong in his judgement of Cézanne's work? If he was prejudiced at all, it was in his favour. We must ask a similar question about the loosely associated group called The Impressionists.

The public reaction to their exhibition was almost hysterical. I cannot help suspecting that this was due not so much to their revolutionary ideas about painting as to their subject matter.

In their paintings, we see laundry women yawning over the ironing board, peasants having a supper that consists of little more than potatoes, and the working people of Paris dancing their cares away.

Cézanne hadn't much sympathy with the working class. He wouldn't have admired the gallant people of Paris who manned and womanned the barricades during The Commune and thus saved the honour of France.

He wasn't around either to disapprove or to admire: he fled from Paris at the advance of the Germans unlike Samuel Beckett who, to his eternal credit, chose to live in wartime France rather than return to Ireland.

The second half of the Nineteenth Century in France — and especially in Paris — was a period of unprecedented ferment in politics and in literature and in the arts.

The first quarter of the Twentieth Century in Ireland — and especially in Dublin — could be compared to it but only in a small way.

Cézanne lived in Paris through some of the most exciting years that the capital ever knew but he seems to have been unaffected.

11

On rare nights he visited some of those cafés where painters and writers and intellectuals casually came together but he always remained on the fringe of the company and hardly ever spoke. He was hopelessly shy; he was in thrall to melancholy and depression and self-doubt but in the midst of all this suffering there was a terrible core of certainty.

Henry Moore paid him the ultimate accolade: "Cézanne battled against all the things that he admired in painting."

Moore went on: "I believe that it is better to attempt something you cannot do rather than do what comes easy."

He exhibited on a few occasions with the Impressionists but eventually felt that he had nothing in common with them and nothing to learn from them — he continued to admire their work, especially that of Edouard Manet and Camille Pissarro.

His method of working became legendary: he took so long at painting flowers and fruit that sometimes the flowers became withered and the fruit became wizened.

He and Gerard Manley Hopkins were more or less contemporaries; they were hardly aware of each other but were remarkably akin.

Hopkins lived in Dublin in yeasty years but he seemed to be unaffected: he had no sympathy for Irish Nationalism; he passed through the city like a ghost and left little trace.

You will find hardly anything of Paris in Cézanne's work; the same is true of Hopkins in relation to Dublin — they were outsiders.

Hopkins believed that the language of poetry had become tired and he laboured to create a new language. Cézanne likewise believed that traditional painting had run its course.

He spent endless hours over what seemed simple themes because he desired to capture what Hopkins called the inscape of things.

It is not an easy word to explain: he meant that things should leap at you out of his poetry in a way that you had never known before.

Cézanne cherished that same feeling: he would love to think that he could paint an apple and make you see it with the eye of childhood, as if you had never seen an apple before.

In his later years he returned to the family home and his first love, the countryside of Provence. He wasn't welcomed by his neighbours. They disapproved of the wealthy man who went around dressed like a tramp and painted in the fields in all weathers. He was letting down the side.

Cézanne's terrible core of certainty sustained him. As he grew older, he became more and more obsessed with capturing the essence of his native place and especially that of the Mont Sainte-Victoire.

DH Lawrence spoke about the messages that came to us from the depths of the earth. It was said that he could describe people in a field and that you would remember the field when you had forgotten the people. Cézanne would have understood.

I understand why the Aborigines of Australia venerate Ayer's Rock and I understand why Cézanne was obsessed with the Mont Sainte-Victoire.

Mountains with their sense of solidity and permanence are like fathers; rivers are like mothers.

The Mont Sainte-Victoire is a friendly mountain; its contours are gentle and its slopes have been cultivated for generations. Cézanne strove to capture its spirit. I believe that his drawings and paintings of that mountain are among his greatest works. While painting in the fields in the October of 1906, Cézanne was caught in a violent rainstorm and on his way home collapsed by the roadside.

He was brought back in a laundry van. On the morrow he got up and worked as intensely as ever. He died from pneumonia a week later.

He wasn't without friends in his later days.

A little group of local young intellectuals idolised him, possibly as much for his heroism as for his painting. In a way, his life had a happy ending.

Evening Press, **Tuesday, November 8th, 1977**

A.E. Coppard — The writer as a sprinter

". . . that, tall man whose height enabled him to look around out of a grave when it was completely dug." When you meet those words, you feel you are in the presence of one of those beings on whom out of an intuitive belief in an ordered universe we tend to confer the label "born writer".

The temptation to think thus of A.E. Coppard is great: so lacking does his youth appear to have been in those nutriments we deem necessary for the fostering of art that the miracle of his achievement seems due to some powerful innate force.

He was born in Folkestone 'the son of George the tailor and Emily the housemaid' in 1878, a time when even that enchanted corner of England was not free from 'Chill Penury.'

George, the tailor, was a radical who 'loved flowers, birds, the open air, and to go a-roving over the hills for mushrooms, blackberries, nuts, cowslips, whatever was reasonable and free, but being doomed, as he was he knew, by tuberculosis, he became careless and something of a drinker and so we were always shockingly poor.'

He died when his son was nine. Then his mother went out to work and the domesticity suffered from her twelve-hour daily absence at a laundry where she had to start out as a plain ironer at 27 pence a day and where she reached the heaven of her ambition when she was promoted to first class at two and six per day.

Young Coppard's academic career ended with his father's death. The nine-year-old entered the world of commerce and was in oil before it was fashionable, daily selling his paraffin from street to street.

That was the first of many trades, including that of professional sprinter; the nearest this short-distance runner came to literature was in the spell he gave as messenger boy for Reuter's.

But whether or not he was a born writer he was assuredly a born reader — and in odd snatches imbibed a knowledge of the English classics that was sharper and sweeter because hungrier than that usually acquired in more orthodox ways.

Almost inevitably he was drawn into the socialist movement; he became secretary of the I.L.P. and went to Oxford — and the famished tree began to put out leaves.

Eventually he had a few stories published in magazines — and the sight of his name in print went to his head. He gave up his job, took a cottage in the country — and set out to confirm the belief of his wife and her mother that he was mad.

And, of course, he was mad — with the madness of a fish going upriver in the breeding season or that of a man who puts his youth in pawn because his ambition is to be a professional footballer.

A.E. Coppard became a professional writer. And if his fame is not commensurate with his talents, it is perhaps because he was that rare being among English authors — one who wrote short stories but not novels.

The short story is a great hunting ground for the mystagogue — so much so that one is tempted to say that there is no such thing.

We tend to believe that because a term exists, there must be an entity to correspond with it — but perhaps all that distinguishes the short story from the novel is its length.

'The Great Gatsby' is an expanded short story; A.E. Coppard's best known short story, 'The Higgler,' is a condensed novel.

Nor does the belief that the Irish have established squatters' rights in this field of English writing withstand disinterested scrutiny — we tend to ignore the practitioners in the other island because almost all were primarily novelists.

Thomas Hardy and D.H. Lawrence and Arnold Bennett and Rudyard Kipling could lead the charge; one might include Somerset Maughan too — he is generally deemed unserious only because he wrote too entertainingly.

Coppard differed from all those in not only that he never attempted the novel — he was a sprinter in prose as he had been on the track — but that he alone wrote with deep knowledge of the rural poor.

Hardy and Lawrence and indeed Kipling knew the fields well enough but not the people who worked them — none of the three had ever lifted a spade in anger.

Coppard's country stories are bone-and-marrow true; they remind one of Crabbe's labourer 'roused in rage and muttering in the morn, to mend the broken hedge with icy thorn.'

But Coppard transcends Crabbe in that his knowledge of rural hardship does not blind him to the glories of the countryside — he sees it steadily and he sees it whole.

Perhaps the most subtle and powerful of all his stories is 'The Field of Mustard': it is a seemingly artless chronicle of a few hours in the life of three women — but its evocations are infinite.

It is a windy afternoon in November and the three are gathering dead branches in a wood — we meet them when they have almost finished and are about to go home.

Rose and Dinah are ready before Amy, 'a slow silent woman' who is preparing a bundle almost too much for her to carry. They go on without her.

When they come to a sheltery hedge, they sit down and wait for her — and in a rare interval of intimacy talk about things that are deep in their minds.

Each envies the other — Rose wishes that she had children; Dinah wishes that she had not.

Both feel trapped by marriage and indeed by life; they are uneducated but articulate — and each feels that there should be far more to living than the fleeting fragments they have known.

Soon Amy comes on; she is overburdened — but also overjoyed because she has found a shilling in the wood. And after supper they will go to her house and share a gallon of stout.

'And again they were quiet, voiceless, and thus in fading light they came to their homes.

'But how windy, dispossessed and ravaged, raved the darkening world. Clouds were borne frantically across the heavens, as if in a rout of battle, and the lovely earth seemed to sigh in grief at some calamity all unknown to man.'

'Dusky Ruth'. Coppard's most famous story, is in a different context not unsimilar in theme: it reminds one of Van Gogh's wall between what you can see and can express — and Sherwood Anderson's 'we all need love but the world has no grand scheme for providing our lovers.'

It begins 'at the close of an April day chilly and wet. The traveller has come to the Cotswolds where though 'the towns are small and sweet and the inns snug, the general habit of the land is bleak and bare.'

That night he is the inn bar's only customer. The girl who tends him is bewitchingly beautiful in a sad way — after almost silent hours they come together like streams in the same catchment.

It is all infinitely erotic but utterly asexual; it is a little story of loneliness and inchoate passion.

In the morning the bar is busy with farmers and others from the world of day; she greets him with curious gaze 'but merrily enough'. After a while he says goodbye and goes his way.

'The Higgler' is also about mute love, but it is a long and ambitious story and a few years ago was made into an unforgettable television film*.

Its central figure is a young man back from the first world war who attempts to make a living as an itinerant rural dealer.

His trade brings him into contact with a remote moorland farm where a mother and her daughter live in seeming loneliness.

The daughter is beautiful and educated but shackled by a combination of pride and extreme shyness; the young dealer loves her but feels her far beyond him.

Too late he learns that she loves him too. It is a tale Hardyesque in its irony — but told with a power of language that Hardy could never command.

'Autumn was advancing, and the apples were down, the bracken dying, the furze out of bloom, and the farm on the moor looked more and more lonely, and most cold . . .'

And then there is this picture of the dealer driving with his horse and cart on a March evening: 'wild it was, though dry, and the wind against them, a vast turmoil of icy air strident and baffling.'

One thinks of Lawrence: it is not so much perhaps that he and Coppard influenced each other but that they were prospectors in the same stream.

Coppard was not confined to the fields: he had known the life of Chaplin's London— and 'The Presser' is a marvellous evocation of it.

It begins; "Two or three years after the first Jubilee of Queen Victoria a small ten-year-old boy might have been seen slouching early every morning along the Mile End Road towards the streets of Whitechapel."

16

And 'when he came to Whitechapel, there was Leman Street and other streets full of shops with funny names over the windows, like Greenbaum, Goldansky, Finesilver, and Artibashev, and shops full of foreign food . . . and women who were drunk at eight o'clock in the morning and who sat on doorsteps with their heads in their hands.'

That the small boy managed to rise above it all and become one of the immortals of literature is a tribute to the resilience and resourcefulness of the human spirit.

A.E. Coppard's writing life brought him what his father had sought with only fleeting success.

"I could believe that summer was always there and I always in the open with the birds and the trees and the postman, the thrift of the land and the freedom."

* *Adapted by Hugh Leonard*

Evening Press, **Tuesday, May 6th, 1980**

Synge — The humble genius

You are never far from the sea in Kerry — physically; mentally, however, it could be a vast distance from many of the county's inhabitants.

They visit it now and then on a Summer Sunday — and their image of it is likely to be sentimental.

Where I grew up was less than ten miles from tidal waters—it might well have been a hundred.

But some of the older men among my neighbours had known the sea at first hand—and they were not sentimental about it.

They were farm labourers who had worked in the rich land immediately to the north of Tralee in the days when seaweed was a great source of manure.

And they had often been out in the "non-human" hours, down to meet the tide with horse and cart — it was hard and sometimes dangerous work.

They were not sentimental about the sea — but that did not prevent some of them from thinking of it romantically: it fed their need for wonder.

I knew one, long settled inland, who loved to walk a little distance from his house to a rise of ground whence he could see the local bay:

He pretended that he did so to help him gauge the weather: if one judged by his predictions, it was not a convincing excuse.

He used to claim that he could hear the voice of the ocean on nights when a storm blew from the west — and he took great pride in identifying the strange birds that sometimes fled inland before it.

17

And some nights too he would walk up the road to the summit of the ridge that looks south over the plain of Mid-Kerry and claim that he had seen the flash of the lighthouse off Loop Head.

John Synge would have understood him: his view of the sea was also unsentimental — but romantic.

Synge was a great traveller, not in the sense of hastening around the globe and seeing the obvious — but of wandering along the byways and perceiving the inscape of people and places.

And in his wanderings he was seldom far from the sea: he seemed to need its proximity — just as some people need a fire in a room even though they may rarely sit near it.

In Wicklow his primal chaos first found a world to which it responded — there he walked in glens and valleys where the sea was only over the eastern ridge.

And in Kerry he felt its nearness even more fully: Kerry for Synge meant mainly the peninsula that extends from the mid-plain to Brandon Point and Slea Head.

There in a sense you are in Ireland's counterpart of Cornwall: you cannot get away from the sea — it is on either side.

And in a sense you are not: Cornwall lies between two great sea-lanes — and you are aware of land to north and south of it; West Kerry is mentally far more isolated.

In Synge's day its sense of isolation was deeper — but even now at certain times of the year it can be overwhelming.

Synge loved West Kerry — in the sense that he felt most himself when lapped around by it and that it aroused in him desires that seemed beyond fulfilment.

And yet the myth persists that he 'found' himself in the Aran Islands after he had met Yeats in Paris and gone there on his advice.

Anybody who has read Synge's logbooks of his travels knows the truth: Aran did not inspire him.

He was affected by its dramatic difference from the world he had previously known — but he sensed there a kind of disillusion that perturbed him.

It seemed like the wisdom of the defeated: sometimes he felt like a child among the islanders — they laughed at him because of his 'romantic notions.'

It was the kind of laughter that Christopher Columbus knew when he announced that he would find a west-ward passage to the Indies.

That is a strange kind of laughter, made up of two elements: one is a sense of superiority over the fool — the other is the fear that the fool might be right.

What perturbed him most of all was to find this deep-seated disillusion in the young.

He knew a girl in her mid-teens who seemed to express the quintessence of that disillusion.

He and she used to have disjointed conversations that tended to peter out with her expressing her view of the world: a town they were discussing would be dismissed—"It's a queer place, and indeed I don't know the place that isn't." People were reduced in the same way.

But this seeming unwillingness to open their hearts did not blind him to the islanders' virtues.

He loved them for their most obvious trait — the courage that was the keystone of their existence.

And he loved too their integrity — that was exemplified in the happenings that gave him the seed of "The Playboy."

Some time before Synge's coming to the islands a man who had killed his father with a spade fled from Connemara to Aran and was protected by the people.

They kept him hidden from the police and scorned the rewards offered — and eventually he got away to America.

But though Synge sets his play in the West of Ireland, its hero is from a distant clime — Christopher Mahon had for eleven long days been walking the world.

When the piranha fish from the academies of the United States have eaten the last shred from Joyce's bones, it is likely that we will see an assault on Synge.

And the significance of young Mahon's name will in itself be material for a thesis.

Possibly Synge intended it — or possibly it suggested itself to him out of the deeper part of his mind and he accepted.

But one thing is certain: the coming of an outsider to create ferment in a community is a perennial element of mythology.

We know him best in this age as the hero in the novel or film of the American West who rides into town and finds himself caught up in its problems.

He stays long enough to bring down the local godfather and put an arrow in the heart of the beautiful daughter of the man whose land the railway company coveted.

Christy Mahon did not exactly ride away into the sunset — but he is a member of the same stable.

And Synge, if pressed, would probably confess that he had a Kerry prototype: the wild poetry of his speech is evidence enough.

'The Playboy of the Western World' has many themes: the most obvious is that of a young man trying to escape from the bonds that threaten to cancel the best part of his nature.

Many an Irishman has to slay not only his da but his ma and a whole drift of neighbours before he can find myself: most do it by emigrating.

The man who leaves Kerry a bashful poor devil and returns a wonder of the western world is not unknown outside fiction — the metamorphosis is, of course, also familiar in mythology.

Christy Mahon does more that that: he starts a fire in the little world where he spends his brief furlough.

A slave to his da he may have been — but there is an almighty store of wonder fermenting within him. He comes to an arid community as the rainmaker.

Consciously or otherwise Synge was paying his tribute to the extreme south-west.

He went there first in 1903 — and returned several times in the few years that were left to him.

His preference for the region above all others is unmistakeable — and there he realised the influence that places have on people.

Light is especially potent—and the light in Kerry has an extraordinary quality.

Synge spoke of it as "wonderfully tender and searching" and wondered "why anyone is left in Dublin or London or Paris when it would be better, one would think, to live in a tent or a hut with this magnificent sea and sky . . ."

But he knew the answer well. There were times when the loneliness of Kerry smote him, the loneliness of a world where life was ebbing.

And he found that loneliness all the more desolating in West Kerry because it was so beautiful.

That kind of world has a curious effect on those who are born there: it arouses a desire to express yourself in some way commensurate with it.

Gaelic football was just beginning to take root in Kerry in Synge's day — and mainly in the towns; music and dancing — and drinking and fighting — were the chief ways in which the people expressed themselves.

And, of course, there were the stories — some of them world-known tales adapted to local circumstances, others created by nameless poets to match the wonder of the sea and the mountains and the fields.

Synge has been called Anglo-Irish: it is a label as wrong-headed as it is sinister.

And, ironically, he would have been more at home at a Munster hurling final than many who deem themselves the purest of Gaels.

The reason was simple: he was a man of total humility — and would have looked on everyone in the crowd as his fellow.

The 'pure Gaels' tend to have aristocratic notions — and when they pine for the mythical paradise that once existed on this island, they picture themselves among its elite.

The much-vaunted Greek civilisation of the golden age had a tiny defect: the perfect democracy rested on the sweaty backs of people who had no voice — the slaves.

A Gaelic state survived until the middle of the last century in South-West Kerry: the area is not yet over its feudal legacy.

Synge at a Munster hurling final would not have been dreaming of Cuchulainn — he would have been observing real people.

And most of all he would have noticed the wandering musicians and the chanters, as the itinerant ballad singers call themselves.

He had an unsentimental affection for those who walked the roads, hardly knowing what the day would bring.

Synge was a wanderer too—but cushioned by a private income; he walked for discovery and contemplation — they travelled so that they might live.

There is no more memorable passage in his log books than his description of a little incident concerning a tramp whose private bit of laundry he accidentally oversaw.

Synge was resting hidden in the heather above a stream when an old man of the roads came along, surveyed his surroundings very carefully to make sure he was not seen, washed his shirt, wrung it out, put it back on, and went jauntily on his way.

The incident was akin to Wordsworth's famous meeting with the leech-gatherer: it indicated the enormous resources of the human spirit.

What perturbed Synge and casts a sadness over so much of his writing is his feeling that the people he came to know had so much to give and were so unfulfilled.

The easy answer is that the native civilisation had been broken, that the country had grown old but not matured.

Synge suggests a more subtle reason — at least for the loneliness he felt in West Kerry. You will find a similar intuition expressed more deliberately in Canon Sheehan's 'The Graves at Kilmorna.'

Part of West Kerry's loneliness came, of course, from the emptiness left by emigration — but emigration went on there even at a time when the fishing was so good that there was an abundance for all.

Was it because that amalgam of sea and sky and rock and earth was so intoxicating that you went forth to find some way of giving your impulses expression?

Christy Mahon was fleeing from his unmurdered da — but he was also pursuing himself.

21

The Races on the Strand

Some think of the Irish nation as a tree: today's people are its present leaves — those past generations have fallen and gone back into the earth and help to nourish the tree.

The image is more attractive than relevant.

And the deeper you question the nature of the concept 'Irish nation,' the more you feel its complexity and elusiveness.

There are peoples who can satisfactorily be called nations — because for a combination of reasons, especially the geographical, they have long been relatively undisturbed.

Ireland, so easily accessible in the days when roads were mainly made of water, was for long a melting pot in varying degrees.

And a curious effect of this is that the people of the west coast — supposedly the 'most Irish' — are not without a mixture that is anything but Gaelic.

But what does it matter? Their ancestors may have been Cromwellian soldiers — but the scape of the sea and sky and land puts its stamp on them.

And the Irish nation of today can hardly be looked on as the successors of the people Brian Boru knew a thousand years ago.

It isn't only that the past is a foreign country — so many waves have washed over the land since then that the blood is well diluted.

And yet we tend to use the term "the people" as if we could define it on demand.

John Synge was one who would question its meaning.

Synge had that rare kind of intelligence: he was as free from the outlook of his time and class as a man could be.

He was also honest with himself: he did not shrink from expressing a view that some people might construe as prejudiced.

And his essay 'A Landlord's Garden in County Wicklow' is a little elegy for a world that he saw passing away in his youth.

The garden was a symbol of that change; it had been left to itself for several years—the apple-trees were covered with moss and lichen, the plum trees were dying, the pear tree was dead.

Synge wrote: "Everyone in Ireland is used to the tragedy that is bound up with the lives of farmers and fishing people — but in this garden one seemed to feel the tragedy of the landlord class also and of the innumerable old families that are dwindling away.

"These owners of land are not much pitied at the present day, or much deserving of pity — and yet one cannot quite forget that they are the descendants of what was . . . a high-spirited and highly cultivated aristocracy."

Out of that class had sprung Grattan's parliament — a poor corrupt thing, it is true, and only vaguely related to democracy but at least a first step in self-government.

And out of that class too came Parnell — who greatly hastened the change that Synge mused over in the ruined garden.

And the day would come not long after Synge's time when many people of that class from which he himself was sprung would feel themselves outsiders in a land that they loved and that their ancestors had diligently cultivated.

The parallel with Zimbabwe today is not exact — but it is relevant.

To those whose emotions need to be fuelled by simplicity the issue is clear: there are natives — and there are Europeans.

And so too there were Irish and Anglo-Irish — the latter term was a dishonest way of saying non-Irish.

It was a simple view — and a potent one.

Yeats's defiant lines that sing the worth of his class could do little to counteract it.

Can you imagine a white Zimbabwean poet who today would pour out his anguish at seeing the world of his fathers in ruins?

Yet some Synge will someday wander in a ruined Zimbabwean garden and wonder where lie the truths of history.

Synge understood Ireland far better than did Pearse — but Pearse's vision, if in a distorted form, has proved far more powerful.

The comparison is hardly fair to Pearse: he had not Synge's leisure — but even if he had, it is doubtful if he would have had his insight.

He was too removed from common reality: he attributed to 'the people' virtues they did not possess — and was unaware of some they did.

But it was hard to blame him: he was caught up in a passionate tide that he himself had helped to create — and passion tends to sweep away complexities.

The concept of the ancient Ireland of Cuchulainn is enchanting—but hardly the best starting point for one setting out to understand modern Ireland.

Pearse's declaration of a Republic was heroically innocent — many a man and woman would see a greater step-forward in the coming of the old age pension.

The great fear of the people in the Ireland of Synge and Pearse was that they would end their days in the work-house.

For many of them their idea of a Republic was frugal independence.

Pearse's nobility and his innocence were manifested in his reaction to the storm that accompanied the birth of 'The Playboy of the Western World.'

He defended Synge and vigorously condemned the attempts to suppress him but he did not like the play: he deemed it a misrepresentation of 'the people.'

What Pearse did not understand was that poverty does not always make people noble — it tends to have the opposite effect.

And the wound left by the Famine is only now closing: part of its legacy was a terrible dread of the spiritual — and an obsession with what for some strange reason was called common sense.

It grotesquely exaggerated the innate conservatism of the peasant — the trait that with cruel irony had been partly responsible for the greatest disaster that befell modern Ireland.

And in the aftermath of the Famine came something utterly foreign to that breed of Irish — loneliness.

The teeming populations of the first half of the nineteenth century had at least the advantage of one another's company.

And the land was cultivated with an intensity that we can hardly imagine today.

You can still see evidence of that cultivation in the most unlikely places — even on steep hillsides now gone back to furze and heather, the corduroy of tillage is visible when the sun is low.

And you will find poor out-of-the-way plots of ground dignified with someone's name because once upon a time he had battled for a living there.

The mass emigrations brought sweeping social change — for many of those left behind it was a devastating experience.

Quite literally they found themselves without neighbours — a great web of social ties had been torn beyond repair.

Old people and widows could find themselves without anyone to do the work on their holdings — their world shrank.

One of the chief spectator sports in the early 20th century was watching the emigrants set out on their long journey.

Synge saw it for himself as he came back one time by train from Kerry to Dublin.

"At several stations girls and boys thronged to get places for Queenstown, leaving parties of old men and women wailing with anguish on the platform.

"Two young men had got into our compartment for a few stations only and they looked on with the greatest satisfaction.

"'Ah,' said one of them, "we do have great sport every Friday and Saturday, seeing the old women howling in the stations.'"

Synge found in Kerry a world that had survived the disasters of the nineteenth century — it still had a warm roof over it and four walls around it.

And especially on the Great Blasket he came on a world that was like a living museum of the Ireland before the great scattering.

The life of the island was then probably at its zenith.

The fishing was good, both for nets and pots.

The islanders had not achieved self-sufficiency — but were near enough to it.

And it was a difficult place in which to be lonely — its people knew something of the rook's security.

They were members of a community in a way that very few in this country are now.

And there was a tenderness that might surprise those who take their view of Ireland from stereotypes.

D.H. Lawrence found a similar kind of concern in Sicily — a land that too has a rather different image.

One aspect of the islanders surprised and puzzled Synge — the men could talk for hardly half-an-hour without mentioning drink.

And the ability to consume large amounts and keep your head was greatly regarded.

Times haven't changed much — the same attitude persists in Kerry today. And it is probably known the world over.

There was a special reason for it in The Blaskets — where men spent so much of their lives on the sea.

But there is a far deeper reason — the effect of drink is often related to a man's image of himself.

The travelling people get drunk very easily — not only because they are undernourished but because they need the release.

And in a tight society such as that of The Blaskets is was believed — if not articulated — that the man who got drunk easily carried within him a dangerous sense of inferiority.

Drink, of course, had another function too — even for the strongest-minded it was an outback.

And on certain occasions there was a general licence to go walkabout. You would be found guilty but sane.

You will find in Synge's travel-log a remarkable piece about the aftermath to Glenbeigh races.

The races there are held on the strand (Christy Mahon, here we come) and the occasion is an unofficial local holiday.

Synge had left the strand after the races — and missed the chief fun.

He is told about it in glorious detail. There was great fighting — and some of the warriors ignored the incoming tide and waged a kind of naval battle.

The parish priest has to wade out to his waist — the warriors battle on. He has to assault them to achieve some degree of peace.

"Another man was left for dead on the road . . . then there was a red-headed fellow had his finger bitten through, and the postman was destroyed entirely."

Evening Press, **Tuesday, July 29th, 1980**

A Poet made by a place and a time

The farm labourers are no longer an endangered species — they are extinct.

There are still farm workers — but the farm labourer in all his glory is no more.

'Glory' may seem a wildly irrelevant word to use in such a context—but it is not.

The good farm labourer — and the profession had little room for bad ones — knew glory, even though it was not of the kind that comes with headlines and cheering crowds.

He experienced it in the inner glow that comes from work well done — even though that work may have been poorly remunerated and carried out in conditions that could be a lot better.

And just as a stable lad may look on a horse as more truly belonging to him than to the owner, so too a farm labourer could regard the fields in which he worked as in a way his own.

It was his spiritual reward—and indeed he had little else: in general, they were an abused class.

The history of the word 'spailpin' is indicative: once it denoted a reaper who travelled around following the harvest; today in those places where English is still speckled with Gaelic it signifies a worthless fellow.

The change the word has suffered is a remarkable tribute to the Irish caste system — because the spailpin, the farm labourer turned specialist for the harvest, was a man of great value.

There is no need to lapse into sentimentality so that he may appear in a good light—the facts are compelling.

The typical farm labourer could, for instance, mow half an Irish acre of hay in a day.

That area is almost four thousand square yards — think about it the next time you are sweating over your lawn.

And he had to know what to do if a horse got the colic or if a cow had difficulty in calving.

26

He had, in short, to be endowed with many skills and varied wisdom — and great hardihood.

And yet you will find little about him in the recorded history of this country.

Indeed, if you wish for glimpses into his life, you must look elsewhere — at, for example, the mowing of the great meadow in 'Anna Karenina' and various scenes in Thomas Hardy.

But there were two Irish farm labourers about whose lives we know a little: one, Owen Roe O'Sullivan, was a poet — the other, Patrick Ledwidge, was the father of one.

Owen Roe's life has been the subject of outrageous romanticism: like Robert Burns he survives in the folk memory as a great rake and a prodigious poet and wit and scholar.

The facts of Burns's life are easily verified: he was a tortured craven man — and his early death was the legacy of rheumatic fever, brought on by working in the cold and the wet.

O'Sullivan's life is shadowier: he was probably a teacher in Winter and a farm labourer for the rest of the year. He spent a while in the British Navy — and died sordidly enough after a fight in a Killarney pub that left him with a head wound that bad nursing compounded.

Patrick Ledwidge had at least two things in common with Burns and O'Sullivan: he died young — and he did not allow the heavy labour of the fields to cripple his spirit.

Poetry had been their anodyne — his was the ambition that his children would get a better start in the world than he did.

Francis Ledwidge in a letter to an American professor wrote: "I am of a family who were ever soldiers and poets . . . I have heard my mother say many times that the Ledwidges were once a great people in the land . . . "

In that touching belief he and she showed themselves to be part of the great Irish dream of their day.

Patrick Ledwidge may indeed have been a member of a family that, like Tess Durbeyfield's, had come down in the world — but whatever the case, he was a migrant labourer when he married Anne Lynch in Slane in 1872.

They settled in a two roomed cottage — and he worked on a farm where between wages and extras such as milk and fuel he earned about fifteen shillings a week. It sounds miserably small—but it was well above the average for the time: you should bear in mind that in the nineteen-forties few farm labourers made thirty shillings a week.

Obviously Patrick was an exceptional worker — and his wife shared his ambitions for their children.

They soon got a cottage from the Rural District Council — at Janeville, about a quarter of a mile from Slane.

It was a spacious little house — you can still see it — and with it was a half-acre of good land.

The Ledwidges' plans were coming on nicely. Patrick, the eldest son, stayed on as a monitor in his national school — the dream was under way.

To read now about Patrick and Anne and their young family is to go back to an age of epic innocence.

It was a time of hope, compounded of many elements—including Victorian self-improvement, Irish nationalism, and the general intuition that the world was improving.

The land was being won — and the dispossessed were confident that they would come back into their little kingdoms.

But the Ledwidges' ladder was shatteringly pulled from beneath them — Patrick died suddenly.

Anne was left with eight children — the youngest was only three months.

It was the kind of tragedy that makes the plight of Shakespeare's protagonists seem self-imposed.

The only state assistance then was a shilling a week for every child — but Anne did not put any of her family into the home.

She was determined to keep them together — and so this mother courage went into the fields to earn what she could by weeding and thinning and snagging and picking.

And when there was little work on the land, she knitted and washed and darned.

The work in the fields was paid by piece-rate — and her prowess was the wonder of the locality.

Patrick had to abandon his hopes of going on to be a teacher; he stayed on at school until he learned book-keeping and got a job in Dublin.

New he became the family bread-earner. But he was soon back home, afflicted with the scourge of the time—tuberculosis.

The young people of today do not know the terrible effects of that disease.

Once the sufferer discovered he had it, he resigned himself to death — and if, as was usually the case, he remained at home, the family were to some extent ostracised.

Patrick lingered on in the cottage near Slane for four years; his mother was again back working in the fields.

One day she faced up to the final humiliation: she could not pay the rent — and the bailiff and the police came to evict the family. The local doctor intervened—he certified that Patrick could not be moved.

His death brought further humiliation — he was buried 'on the parish'. Francis later wrote of the years when it seemed 'as though God had forgotten us'.

There is a strange parallel between the life of the Ledwidges and that of D.H. Lawrence's family.

You will remember how Lawrence's mother was determined that her eldest son would not go down into the mines — and kept him at school until he got a job as a clerk in London.

He too was soon dead — and the Lawrences knew the dreadful feeling that comes when you believe that fate has turned against you.

Anne Ledwidge seemingly believed that God moved in very mysterious ways — at any rate, Francis never heard her complain. She was free from the worst of all Irish diseases—self-pity.

The dream though tattered went on. The children were kept at school until they could find reasonable places in the world.

Francis left school after his Confirmation. He was almost fourteen — it was not uncommon then for lads of ten to be full-time labourers in the fields.

Young Ledwidge became an apprentice cook at Slane Castle. But his stay was short — his sense of humour undid him.

One morning he rubbed out the day's menu on the kitchen slate — and substituted such delicacies as spuds and bacon and cabbage. The cook was not amused.

He next tried the grocery trade — and spent three whole days in a shop in Rathfarnham. Then he stole away in the night and walked the thirty miles back to Slane.

Although he was not fully aware of it then, he had made a momentous decision: he had declared himself a poet.

Since childhood he had been uttering and scribbling verses — and in his Dublin exile had suddenly felt that poetry was his vocation.

And he sensed too a truth that a great Greek poet, Cavafy, was later to express: "In those few streets or fields where you grew up, there you will live and there you will die."

Back home he got a job as gardener-cum-yardboy with a liberal young couple who treated him in a civilised manner and paid him far above the average wage.

After three years they left for Dublin — and Francis, now twenty, went to work as a road-mender with the county council.

In between jobs he had often worked in the fields — and in his fumbling way was becoming well equipped to understand his heartland.

It is often said that his poetry is too literary — that he wrote out of The Golden Treasury rather than out of life.

But that is not true at all—and those who say so do not know Meath, that county of rich land and noble trees and deep canopied roads and a great sense that nature is bountiful.

And if the stony grey soil of Monaghan burgled Patrick Kavanagh's bank of youth, so did the essence of Meath affect Francis Ledwidge.

There was another great difference between them: they grew up in eras that though separated by only a generation seem centuries apart.

Ledwidge's Ireland was full of yeast: 'Ulysses' is a brilliant book—but it performed a post-mortem on a body that was about to waken from a deep slumber.

Ledwidge grew up in an atmosphere of hope that now seems pathetically naive—but that was, in fact, sensible and wholesome.

He studied the Gaelic language in his spare time; he played the new code of football; he was a vigorous worker in the budding labour movement.

Kavanagh's Ireland gave out a sour smell like an abandoned porter jar.

He grew up in a climate of economic depression and spiritual sickness.

The sickness came from a sense of betrayal: 'native government' (was it ever really such?) had not brought the promised land.

The miracle of Kavanagh is that he sang so well in such a place and time.

And if there is a softness in Francis Ledwidge, it comes from two things.

He lived in a world of an innocence we have lost forever — and his poetry was written in youth.

He went across the river and into the trees at thirty.

Evening Press, Tuesday, August 12th, 1980

A Poet of magic fragments

One of the myths attached to Francis Ledwidge is that he was 'a gentle dreamer.'

'Dreamer' is a much-abused word: in this context it implies a degree of unawareness of life's harder realities.

In that sense Ledwidge was not a dreamer—but he was a dreamer in that because he was acutely aware of what life was he dreamed about what it might be.

And if he was gentle, it was not in the sense that he shrank from the rough and the rude.

One Sunday evening, for instance, after his Gaelic football team, Slane Blues, had been beaten by Navan Harps, he did not turn the other cheek.

He was all alone when some of the Navan supporters made uncomplimentary remarks to him about the Blues. Ledwidge hit first, hit again afterwards, and provoked a minor riot.

Later in a letter to the Drogheda Independent he wrote: "Navan Harps should only be allowed to play football in their bare feet and with blocks on their necks like vicious dogs."

Poets in the popular image are frail helpless poor devils—Ledwidge hardly qualified: he was a hardy man who was good at the high jump and thought nothing of cycling forty miles a day in all weathers when he worked as a road ganger.

And dreamer that he was, he sensed that the condition of the workers in his day was not necessarily part of a fixed order.

He was only a little over twenty when he organised something almost unknown in Ireland then—a strike.

He was working in the copper mine in Beauparc with about a hundred others—in conditions that were bad even by the standards of the day.

Flooding was the biggest evil—the men often worked drenched to the skin. Fall-ins were another — death and serious injury were common.

The miners asked Ledwidge to lead them. He did—the strike began. He was instantly sacked — his 'comrades' returned to the mud and the drenchings and the danger. He went back to work on the roads.

He could hardly be blamed for having overestimated their spirit: that first decade of the century was a time of yeast — an exaggerated concept of Ireland was abroad.

The Gaelic League was a manifestation. Ledwidge attempted to start a branch in Slane.

But the organiser for that part of the county seems to have been one of those intellectual snobs who ensured the failure of the movement — Ledwidge was spurned and attempted to learn the language on his own.

It was a perplexing time for the young idealist: he was overflowing with awareness of the forces stirring below the quiet surface of Meath—but now he had twice failed to convert spirit into practical reality.

He was soon to know a third failure—and it went far deeper.

He was walking out with Ellie, the girl who was to him rather as the Miriam of 'Sons And Lovers' was to D.H. Lawrence.

Their courtship gave rise to the little poem that begins:

I feel that she will come in blue
With yellow on her hair, and two curls strayed
Out of her comb's loose stocks . .

As in all tales of true love there was an obstacle — but in this case it was not to be surmounted: she was the daughter of a strong farmer—he was a man of no property.

That the courtship ever started was remarkable; Ellie would have needed a heroic strength of character to continue it.

Her family liked Ledwidge well enough—but he would have had a better chance of marrying a lord's daughter. Even today the Irish rural caste system permits few exceptions.

The poor girl's elders besieged her with the deadliest weapon in Irish life — common sense. She yielded—and broke the news to Ledwidge.

He was desolate — but a poet carries within him a great anodyne: he can lighten the heaviest grief by trying to cast it into words.

One of his poems about that crisis begins:

You looked as sad as an eclipsed moon
Above the sheaves of harvest . . .

Another ends:

And now that you are lost I may pursue
A sad life below the depth of words.

It is an enigmatic statement: it is clear, however, from the body of the little poem that Ledwidge sensed he might find in poetry a compensation for all his rebuffs.

He believed at the same time that Ellie might return to him—but she married one of her own class. By then Ledwidge was in the army—and broken love was one of the complexity of causes that took him into it.

Another of the myths attached to him ascribe his joining-up to the advice of his Meath neighbour, Lord Dunsany.

It is a silly slander. Ledwidge was a remarkably strong-minded young man — and he and Dunsany met as equals.

Dunsany never in the slightest sense patronised Ledwidge. He encouraged him, helped with grammar and punctuation, and got him a publisher.

He also envied him — because the son of the farm labourer had inherited a vitality of language unknown to Dunsany's class.

The loss of Ellie was probably the happening that pushed Ledwidge into the army—but it would be wrong to ignore less personal reasons.

Ledwidge—as you will see from his letters—was a nationalist but hardly a separatist: it was a stance he shared with some of the Fenians.

Perhaps he sensed that socialism was more likely in the context of the two islands rather than in feudal Ireland.

Perhaps his own bitter experience with Irish values had put a question mark behind the word 'nation.'

There was another factor: Germany was then looked on as the world's greatest military power — after all, the Franco-Prussian war was a recent memory — and Ledwidge had sense enough to know that defeat for Britain would not profit Ireland.

And, as he said, he did not want other people to do his fighting for him. And, paradoxically, he was one of the few Meath Volunteers to oppose John Redmond after Woodenbridge — and yet he enlisted.

Only three years of life remained to him.

And, as a man will when he realises his time may be short, Ledwidge lived those years intensely.

Another factor deepened him:

Ellie's marriage did not turn out well; her will to live weakened—and she died at the birth of her first child.

Ledwidge now felt that they were nearer to each other than ever — it was a kind of Wuthering Heights love.

If suffering helps to make a poet, he was growing fast. And the Rising of 1916 added to his spiritual turmoil.

A famous Irish journalist has told of meeting Ledwidge in Manchester in Easter Week; the poet was on his way to join the insurgents—in Dublin he could not make contact and came back.

Thus arose another Ledwidge myth. The truth is that like many other Irishmen in the British forces he was tortured by his dual allegiance — but he stayed in the army.

He found some solace in writing a tribute to Thomas MacDonagh — it became his best-known poem.

He was also court martialled for 'insubordinate talk' about the Rising.

And despite his heroic service in Gallipoli and Serbia, he lost his corporal's stripes. But he hardly cared. He was sick of war — and lived now only for his poetry.

How good is that poetry? It has long been fashionable to describe him as a minor poet—but that is a vulgar judgement.

To classify poets as if they were football teams in a league is as wrong-headed as saying that an oak-tree is better than a rose-bush.

John Clare, for instance, created nothing as grand as 'Paradise Lost'—but some of his scattered utterances pierce deeper than anything in Milton.

Francis Ledwidge's output was not huge—for most of his short life he was busy making a living or toiling in the many streams of what you might call the Irish Renaissance.

And strong-minded though he was,, he knew his formal education was only rudimentary; he could not but be somewhat in awe of the established poets.

He knew the feelings of Thomas Wolfe, the young man from the mountains of Carolina, waking up in his lodgings in Chelsea and wondering how he could hope to walk in the fields of the great.

For a while Ledwidge was like a peasant who thinks that the songs from the great world outside must be better than his own — and he imitated them.

But even in the early poems the borrowed dress does nor altogether mask his originality: always there are outcroppings that are pure Ledwidge—and pure gold.

Hardly a single poem of his is perfect — in the sense that Hardy's 'Weathers' and Hopkins's 'Felix Randall' are: he is a poet of memorable fragments.

In 'Twilight in Middle March'

A gypsy lit a fire and made a sound
Of moving tins . . .

You remember the magic of those simple words.
'Autumn' begins

Now leafy winds are blowing cold,
And South by West, the sun goes down,
A quiet huddles up the fold
In sheltered corner of the brown.

That, you may say, is very much in the English lyrical tradition. So it is— but its language is fresh and bold.

And there is:

The sheep are coming home in Greece,
Hark the bells on every hill,!
Flock by flock, and fleece by fleece,
Wandering wide a little piece.
Thro' the evening red and still,
Stopping where the pathways cease,
Cropping with a hurried will.

You are there.

Perhaps there is one poem of his that is so near perfection that you can hardly quibble.

It is a great favourite of the anthologists—and its last stanza has been quoted as often as Patrick Kavanagh's lines about the stony grey soil of Monaghan.

Ay soon the swallows will be flying south,
The wind wheel north to gather in the snow.
Even the roses spilt on youth's red mouth
Will soon blow down the road all roses go.

The early part of that poem 'June' is often criticised as being too much influenced by John Keats.

Perhaps it is—but it could be less influence than kinship.

And if Ledwidge's world seems to lack the harsher realities of Kavanagh's, it is possibly largely due to the influence of place.

He might own nothing of Meath—but it was his world: its beauty and sense of bountifulness were great elements in shaping his soul.

And even if he had lived to a ripe age, it is hard to imagine him as being other than lyrical.

He dedicated his book of poems: "To my mother, the first singer I knew". Take away his urge to sing—and he would cease to be a poet.

But singers are not necessarily blind to the world's sorrows—and one line in a poem about that same woman captures a subtle and powerful element of Irish life, especially in his time.

'For there is that in her which always mourns.'

On the 31st of July, 1917 while he and his comrades were drinking tea in the country near Ypres, the poet of the Royal County was killed instantly by a shell.

He was a few weeks short of thirty — but already famous.

His fame eased the pain for the mother who had laboured so heroically in the fields to keep her family together after her husband died.

It was a consolation to have reared an immortal.

Evening Press, **Tuesday, October 6th, 1981**

O Conaire's quest for himself

There is a tendency to attribute famous sayings to famous people but perhaps we are safe in believing that Samuel Johnson said: "When a man is tired of London, he is tired of life."

But before we rush to agree with him, we might remember that the London of his day was not the great conurbation we know now.

It was a small city, built mainly along by the river — and when the denizens of the Fleet Street coffee-houses proposed a walk in the country, they had in mind such remote and untamed places as Paddington and Chalk Farm and Shepherd's Bush.

And another indicator of London's smallness a few hundred years ago is the supposed origin of the word 'Soho'.

The story goes — and it may be true — that once upon a time the territory now known by that name was a marshy flat.

Men used to course hares there — and as everyone knows, when you spot a hare in his form and wish to catch him alive, you try to hypnotise him by saying "so-ho, so-ho — so-ho" and so-ho on.

And not only was Johnson's London small — it was dirty and ill-lit and dangerous.

The word 'mugger' was hardly known then but the word 'footpad' was — and when you were hit smartly over the head and woke up to find yourself robbed, the trade-description of your assailant hardly mattered.

The emergent city by the Thames offered other hazards: as you went about your business, you stood a reasonable chance of having a chamber-pot emptied over your head — and if you drank too much coffee, you might stagger into an open sewer on your way home.

Yet Johnson, that seemingly-solid son of the midlands, loved the place and felt homesick when out of it.

If he had persevered in his studies, he could have got a secure and not uncongenial post in Oxford or Cambridge — he chose London and suffered the nadirs of misery and humiliation before he reached safe harbour.

It is not too hard to understand why: London was then the capital of the world ("where it was all happening" we would say now) — and Johnson was drawn to it as a dog is to the fireside.

You cannot measure spiritual expansion — but you can say with certainty that if William Shakespeare had spent his life in the west country, he would have been a lesser artist.

He came to London — and knew it when it was the nerve-centre of a young country spreading itself on the waters of the world.

The energy and exuberance of his work came from the air all around him — England was like an adolescent becoming aware of his muscles and sensing that they were powerful.

It is hard to say where the world's capital is now. Things are so strange that it could be in Tokyo or even in Los Angeles — more likely it is nowhere. Certainly there is no counterpart of Athens and Rome in their heydays.

Samuel Johnson had no doubt about its location: it lay in the few miles along by the river from Lambeth to Southwark.

And like Shakespeare he lived in an era of expansion: the exploitation of the East was reflected in the crowded Thames.

And as happened in Athens and Rome and Paris, the capital of the empire throve on the riches of the colonies.

At least the ruling class throve: for the mass of the people imperial glory meant unemployment at home plus the chance of dying in battle abroad.

Yet so pervasive is the tribal instinct that those who rejoice most in their country's military exploits are those who gain the least from them.

Beranger's poem about the old peasant woman who venerated Napoleon expresses that paradox well.

But for the Londoners (and the British in general) there was an intangible benefit: military success engendered faith in themselves as a people.

You might argue that it was faith in themselves that brought the military success — but Britain's armies, like those of most empires, were not recruited entirely from the mainland.

This aspect can be conveniently forgotten by the myth-makers — but the myths are none the less potent.

And the confidence of the British — and especially the English — is related to the myth of their military invincibility.

And, of course, that invincibility was related to their prowess on the sea.

London is not lacking in symbols evoking memories of glory on land and water.

And though the ordinary Londoner has cause for being sceptical about such glory, he is proud of it.

Hitler's war added a gloss to the myth: "London can take it." It wasn't untrue — but some German cities took even more.

The effect of all this truth and near-truth is a city from which emanates a sense of fulfilment.

In Solzhenitsyn's "August 1914" there is a memorable passage about the impact of East Prussia on the Russian soldiers that have entered it.

They are poor peasants used to living from day to day. The fine houses and the sense of wealth and ease make them feel they have come not to a new country but to a new planet.

Many an Irishman coming to London for the first time has known a similar feeling.

The spirit of a place is intangible — but it is real.

It is impossible to define—but impossible not to feel.

D.H. Lawrence used to speak of messages that come from the depths of the earth — and in "Kangaroo" he marvellously invokes the spirit of Cornwall.

To him it was a land of dreams, haunted by its past and uncertain of its present.

London is a city of the here and now, aware of its past but living in a fertile present.

And the awareness of the past affirms the present — because that past is so rich in success.

To the Irishman even the old yellow-brick streets of Kilburn can seem a promised land.

And the inner city has a majesty that feeds his greatest hunger.

And London has another dimension for him — in its air is a great sense of freedom.

And it is something of a paradox that while Britain's empire-builders were pressing on to put more and more red patches on the map, the seeds of democracy were growing steadily at home.

Karl Marx burrowing away in the British Museum is a symbol of the tolerance that is one of the roots of democracy.

To the Irishman it gives a freedom other than political — it is the outback to which he may discover himself.

At home he is burdened by the peculiar Irish albatross — his awareness of how his society sees him.

He senses in himself something not contained in his society's image — and hopes to find it in the exile that may be his spiritual home.

George Bernard Shaw expressed this dilemma rather hysterically in saying that if he ever returned to Dublin, he would curse every single stone of it.

And yet it is strange that London, the spiritual home of so many writers, is itself so little expressed.

It never found its Zola or its Simenon. And perhaps it is best evoked in a novel written in Gaelic by a young man from Galway.

Padraic O Conaire's 'Deoraiocht' could in another sense be called the great Irish novel — because perhaps it comes nearest to capturing what one might call the nucleus of the Irish soul.

And O Conaire's own career exemplifies almost too neatly the importance of London to an Irishman.

It is so neat that you suspect it happened because he believed so much in it — but for whatever reason, almost all his worthwhile writing was done in London.

His return to his own country was rather like that of Oisin from The Land of the Ever-Young: when he touched the soil of Ireland, his magic deserted him.

O Conaire was born in the city of Galway, spent much of his childhood in Connemara, and — like many sons of the Irish middle-class in his day — joined the British Civil Service.

He was posted to London, was caught up in an intellectual ferment that included involvement with a branch of the nascent Gaelic League, and 'discovered himself' as a writer.

Some eminent critics have said that 'Deoraiocht' was Russian-inspired — and perhaps it was.

Whatever the source of the impulse, the book has a powerful life of its own.

It is as relentless an exercise in realism as anything in Zola — and yet it leaves you with a great sense of exhilaration.

Unconsciously or otherwise O Conaire exemplified a famous dictum — that a work of art no matter how depressing the life it portrayed should open the soul.

It is impossible to read 'Deoraiocht' and not feel the better for it.

It is impossible too not to suspect that it is more than a straight narrative — it is an epic parable.

It begins in London where its hero (or anti-hero before his time) is injured in a traffic accident.

He loses an arm and a leg, gets compensation money, and squanders it.

This theme recurs in his best short story, 'Paidin Mháire'.

O Conaire sees squandering as something to which the Irish are peculiarly prone — or at least a certain kind of Irish.

They are the people whose poetry is in their living, whose approach to the world makes the charge of the Light Brigade seem to have been meticulously planned.

"Deoraiocht' is partly about the clash between them and those whose prudence is so extreme that it constitutes a negation of life.

The battle is familiar — but especially bitter in its Irish context.

The hero returns home — and learns that a maimed and penniless young man is hardly reckoned to be among the country's more eligible bachelors.

He endures dreadful humiliation — but somehow retains his buoyancy of spirit. He goes back to London — and to a fate that had long seemed inevitable.

Without being over-pretentious you might say that 'Deoraiocht' has three elements that make it great.

It expresses fiercely a big truth about Ireland — that the dominant group have the Puritan vices but not the Puritan virtues.

Another element is its sense of the indestructibility of the spirit — while there is hope, there is life.

And perhaps the biggest element of all is its sense of the life of a great city.

O Conaire does not set out to 'capture' London as Zola captures Paris in 'L'Assommoir' — but its teeming diversified life is powerfully evoked.

And you feel that its scope for adventure — spiritual or physical — is infinite. Samuel Johnson's oft-quoted concept is buttressed.

You are made aware too that not only do Irishmen find themselves there—some are lost overboard too.

But, above all, the mental climate there is more conducive to the life of the spirit than that in Ireland — and in this finding the popular image is inverted.

The maimed hero finds in London someone who loves him — one need not be a symbol hunter to see that as more than an accidental happening.

And you are reminded of another maimed hero, Jake Barnes in 'The Sun Also Rises' — and you wonder how much is he a symbol of Ernest Hemingway's own mental trauma.

For reasons that are not clear O Conaire left the Civil Service in 1914. He was 33—and it is likely that he wished to become a full-time writer.

And because Ireland seemed his orchard, he returned home. The sequel was hardly surprising — he became a 'character' rather than a writer.

He died in the Richmond Hospital in Dublin in the Autumn of 1928, probably of pneumonia brought on by malnutrition.

It is part of Irish folklore that his worldly goods were placed on the table by his bed—a few ounces of tobacco, a pipe, and an apple.

Albert Power's famous statue seems to express a wise old man: Padraic O Conaire had wisdom but he was not wise — and when he died, he was forty-seven.

Evening Press, Tuesday, March 11th, 1980

America, O America

The opening passages of 'A Moveable Feast' are very beautiful: you remember especially the goatherd bringing his little flock along the early morning streets of working-class Paris and delivering the milk straight from udder to can.

You feel that you are entering a treasure land — alas, the book rapidly degenerates into a desert of boasting and calumny and falsehood.

It is symbolic of Ernest Hemingway's own life: the early work, even when derivative, is sharp and fresh and exciting — the 'great' writer became a cruel parody of the hungry young fighter.

'A Farewell to Arms' is naive and pretentious and in parts sham-lyrical — but it is exciting: you are face to face with a writer who loves his craft.

'For Whom the Bell Tolls' is a ludicrous piece of work — if one can apply the word 'work' to a product in the making of which the author hardly broke mental sweat.

It might, without a trace of irony, be called "Oh What A Lovely War": the filth and hunger and the obscenity and the fear and the boredom are all washed away — and in their place you have the mountains and the pines and rabbit stewing over the wood fire and the lovely Maria waiting in the sleeping bag.

But it is a book not without significance: its hero, Robert Jordan, is an arch-Puritan — he is among Hemingway's elite of the saved. And, of course, he has a counterpart — the guerrilla leader who defects when he imagines himself a member of the better-off classes.

Robert Jordan is hardly a newcomer to the world of fiction: indeed, he has been long familiar — he is the hero of a multitude of English schoolboy stories.

He is the one who athletically and morally dwells in a plateau high above the generality; he is the perfect knight who shoots the winning goal or races over for the decisive try in the cup final.

Robert Jordan is only superficially different: he takes the girl instead of the cup. And he is killed in the last chapter — it was an extremely fashionable ending for young men who fought on the side of the 'good' in the Spanish Civil War.

And almost always in Hemingway the world is divided into the good and the bad — not in theological terms but according to the male code: there are those with whom you would go into the desert — and those in whose company you would feel unhappy if you were only hunting jack-rabbits.

That is one of the themes in 'The Sun Also Rises': Cohen, the wealthy Jew who pursues Brett, can never be part of the circle of which she and Jake are the centre — morally he is an outsider.

Even 'The Old Man and the Sea', a book with only two human protagonists, is not free from this snobbery: at the end the veteran fisherman, the brave un-self-pitying stoic, is canonised by contrasting him with the tourists who come down to the wharf to see the ruined giant marlin.

Yet Hemingway remains big — and peculiarly American. Would any European writer have been so obsessed with 'proving' that he was free from even the faintest taint of the decadence that is often associated with artists?

It is not perhaps insignificant that he invariably referred to writing as 'working'. And when he spoke of being better than Shakespeare or Tolstoy or whoever, it was in terms of the boxing ring.

And isn't it likely that his enormous popularity in America was due in part to the code in which he pretended to believe? Essentially it differs little from that of Teddy Roosevelt.

Man's depths were tested in circumstances where nerve and skill stood between you and violent death—perhaps in war or facing a lion in the long grass of Africa or battling with the demons of the sea.

Hemingway appealed to the frontiersman that lurked in every American male long after the last covered wagon had rolled west and the last Indian had snared trout in waters that no white man had ever seen.

If he had believed in reincarnation — but backwards in time rather than forwards — it is likely that he would have nominated himself for the part of Kit Carson, the arch-American hero of his boyhood.

Carson had done what boys dream about: he was trapper and explorer and ceaseless adventurer—he was Hemingway's man who lives his life 'all the way up'.

And he had not only been the greatest of that breed but the last: he was still alive when Hemingway's father was born — he had known the American frontier while it still had secrets to yield.

And it is perhaps significant that the Americans still use the word 'hunting', even when the quarry are half-tame deer: it evokes the pioneer whose skill and lore were big elements in survival.

And that mental remnant of the frontier is perhaps the main reason for a peculiar characteristic of American literature: few of its big writers deemed sport beneath their ken.

Some seem as if they are writing a statutory piece—in a manner akin to the flower passages in English poetry.

Others write out of enduring passion: Sherwood Anderson is a prime example — his love of horses was one of the factors that compelled him to be an artist.

And others — men of high talent — were so in love with sport (some would say 'besotted') that they wrote of hardly anything else. Grantland Rice is an obvious example.

The pragmatist might say that it is a question of demand and supply and monetary reward — it is hardly so simple.

The thread of sport runs so strongly through American writing that it surely represents something in the collective consciousness.

Behind Ernest Hemingway's endless posturing as boxer and hunter and fisherman and aficionado of the bull ring and the rest lies a small boy's hunger for approval.

If he had been a good footballer or pugilist (despite all the talk he hadn't even one registered fight in his life), the total man in him might have been set free.

Fate played another trick on him: he became comparatively rich at an age when many writers are still at least pretending to be living on brown bread and onion soup in a garret.

That, too, is a peculiarly American enemy of promise: far more writers have been diminished by affluence than have withered away from poverty and lack of recognition.

We all know Sean O'Casey's words about a few pounds in the pocket being good for the nerves. And Juvenal had a point when he said that a poet finds it hard to concentrate on his trade when the wind is whistling a tune through a hole in his breeches.

But the fox did not acquire his awareness on a regularly-sated belly.

American writers run a fearsome gauntlet: a best seller in that sub-continent can set you up for life; the land bristles with foundations waiting to smite the unwary with grants; rich women roam the cocktail belt with savage intent — artists have long been fashionable prey.

The writer who blasts off in a shower of brilliance and then loses course and levels off as an opulent mediocrity is as much a part of the American scene as the academic whizz kid who ends up as an executive in television.

Hemingway did not surmount his fame and his money: he lost much of his birthright and became a kind of perpetual tourist.

He reminds one of a character in a novel of Thomas Wolfe's who was always talking about going to Spain 'to do a little writing',

Hemingway wouldn't have said anything so silly — but he had a fatal lust for writing about the fashionable: he was a kind of literary jet-setter.

The running of the bulls at Pamplona is no doubt a very fine institution—one would not mind his writing about it if it hadn't blinded him to the raw reality of his own country.

Wolfe loved the Continent too — but the best of his work sprang from his own country and in an age when America was face to face with its frightening economic vulnerability.

You remember Wolfe's Montmartre and Montparnasse — but you remember more deeply his homeless people huddled in public lavatories in a savage New York winter during the Depression.

And you remember too the mad blind tribal hatred that he had known in his native Carolina: "when they kill a negro in my town, they kill him hard."

Wolfe did not believe in going down to Spain to do a little writing: he felt that you could write just as well, if not better, in the Bronx or Dayton, Ohio, or Hoboken, New Jersey.

One of his novels begins with a sentence that is almost an affirmation of this belief: "The tragic light of evening falls on the rusty jungle that is South Brooklyn."

In his determination to dig in native grounds he was in a good tradition: he was following Henry Thoreau and Mark Twain and Stephen Crane and Edgar Lee Masters and Sherwood Anderson and all those who believed that American writing should be wrought out of the local life.

It was a praiseworthy ambition — but an alarming one: Wolfe tells how he used to wake in his lodgings in Chelsea and listen to the policeman on his beat and wonder at his own temerity in attempting to compete with the great writers who had come from such a mature world.

Could you find or lose a paradise in a small unformed hill-town in Carolina? Could Dante see Beatrice in Atlantic City?

Wolfe pretends that he agonised over the question—he well knew the answer: it had been provided by Thoreau, among others.

And Mark Twain had put the multitudinous life of the Mississippi into words — and it was more exciting than Seine or Thames.

And those doughty pioneers had the sense to write in the language that had been moulded in the new continent.

The written word, of course, differs greatly from the spoken: it is a kind of distillation — but the smell and taste of America is in the language of those early giants.

And do not blame Walt Whitman if at times he is turgid almost beyond comprehension.

Like Wolfe, he was sometimes so overcome by the world around him that he became a hurler of words that did not express his impulses.

That is a dilemma common to many of the early American writers — you will find it in such comparatively forgotten frontiersmen as Ed Howe and Hamlin Garland.

And you sometimes wish you would find it now — the typical modern American writer is never less than polished.

John Updike, I suppose, is a fair sample of the new breed — 'Couples' is possibly his idea of the great American novel.

It is a heap of bright rubbish — but it is not without significance: the new frontier is sex — or what passes for it.

Updike wasn't a bad writer one time — but he, too, was seduced down the primrose path. Literature has taken over from life — he is unlikely to return.

But artists, more than most, are affected by the spiritual weather — and the mental climate of modern America is far more likely to produce a satire than an epic.

The most powerful nation in history has lost its way: Vietnam and Nixon are only more visible wounds.

And Carter's effusions on Afghanistan are about as far as you can go into the jungle of hypocrisy.

Annexation of neighbouring territory is a hallowed American practice. Once upon a time it was done openly—now it is done with a little subtlety, as in Chile.

And yet Carter, looking as if Mom's apple-pie wouldn't melt in his mouth, can go on television and advocate a boycott of the Moscow Olympics.

America needs a Swift or a Voltaire — but too many of its writers are funded by fellowships from institutions that made their money out of human misery.

And yet one feels that it is only a hiatus — that Thomas Wolfe was right when he said: "We are lost but we will be found."

Thoreau and Mark Twain and Sherwood Anderson foresaw the American tragedy—foresaw that the money ethic would bring the great new world to moral ashes.

And yet they were also men of infinite hope. And perhaps there is no more powerful symbol in American writing than the last passage in 'Winesburg, Ohio'.

It is morning and the train is about to leave a little Ohio town for the big city. On board is a young writer—going forth to conquer the world.

Evening Press, Tuesday, April 24th, 1977

And you, Thomas Hardy . . .

They said that Francis Ledwidge wrote out of John Keats, and perhaps he did.

And Patrick Kavanagh used to say that all the Irish poets of this century wrote out of The Golden Treasury.

He meant, of course, all the Irish poets except himself, and perhaps he was right.

But is there a better handbook? If you thought you had a fledgling poet in your household, should you hide Palgrave's cornucopia away from him?

And what did Kavanagh himself write out of? Did the stony grey soil of Monaghan really burgle his bank of youth?

If the family had been given a Land Commission farm in royal Meath when he was a wee lad, is it possible that he would never have uttered a line at all?

Heaven knows Kavanagh had little excuse for being a poet—he was big and strong and could play football and hold his own in rural sniping.

He wasn't like Pope who spoke too truly of "this long disease, my life" or like Byron who overcompensated grotesquely for the thorn in his flesh.

And why were you a poet, Thomas Hardy? You were trained to be an architect—and that was a great step forward for a stonemason's son.

And your mother and most of her ancestors came from that anonymous army whose toil for centuries unnumbered underpinned the world.

They brought in the cows for milking in the pre-human hours before the dawn.

They knew the long heavy days of harvesting when the sun was the clock and seemed to be stuck in the sky.

And they dug and they sowed and they reaped and they hedged and they scrubbed and they washed and they cooked and they baked — and were deemed labourers and serving women and social untouchables.

Wouldn't your mother have been the proud woman if you had followed your profession and prospered (as you surely would — being so mean and energetic) and be known as Thomas Hardy, the architect?

But you didn't, even though you were tight-fisted and a social climber, you chose the briary path.

Why, Thomas Hardy? It wasn't that your ambition was to become rich out of being a professional novelist. When the money came, you were surprised.

Perhaps you have given the answer yourself in that almost-forgotten poem, 'The Peasant's Confession'.

One day you were reading Thiers's Historie de l'Empire—and a little passage in it planted a seed in your mind.

It told how the Battle of Waterloo would almost certainly have been won by Napoleon but for a seemingly-small mishap. An officer had been sent to find Grouchy and give him the order to move so that he could prevent the intervention of the Prussians.

The officer and his message never arrived. What had happened, nobody on the French side ever found out.

But your imagination, Thomas Hardy, out of that passage created a simple but powerful poem.

You see Grouchy and his thirty thousand men slowly passing a peasant's cottage 'deeply in a vale recessed'.

It is mid June — and the peasant is troubled that the war will spill onto his crops.

That evening the fateful officer rides up to his door and asks which way had Grouchy gone. The peasant leads him—astray.

After a while the officer suspects the treason and draws his pistol. The peasant draws the officer's sabre — and strikes.

I hid him deep in nodding rye and oat—
His shroud green stalks and loam;
His requiem the corn-blade's husky note—
And then I hastened home.

The peasant's stock and crops remain intact—but not his mind. At heart he was for Napoleon — and thenceforth he had to live with rue.

And you, Thomas Hardy, were for Napoleon, too; you were glad that his men had never come to Ludworth Cove — but you understood him and you were akin to him.

You had a magnificent talent—no, we will not compromise: you were a genius—and you knew it.

And the genius would not be fulfilled by repairing churches—or even by creating cathedrals.

You were a wordman, Thomas, and it is easy to understand why: you were a delicate child and you were reared very close to the apron strings.

Your body grew slowly, as if apprehensively feeling its way in a hostile world—your imagination put on wings and was your Pegasus.

And there in the kitchen amidst the smell of baking and boiling and stewing your mind was filled with wonder.

Your mother told you tales that were at once melodramatic and true: it was an age when people were desperate—and when the power elite looked on hanging as both an excellent corrective and a popular sport to placate the masses.

And sometimes for some task that needs many hands, such as filling the puddings after a pig-killing, some of your mother's friends would be in and you would imbibe their conversation.

They would hardly notice you as they talked about the different farms they had worked in, about the drudgery and the humiliation, but about the sweetness, too.

And the egg of Tess was dropped into the waters of your imagination — and the sand covered it and it grew into a beautiful fish.

And Tess was, of course, your own mother, Thomas Hardy—no American scholar, no matter how voluminous his clues and no matter how sophisticated his battery of computers, will make that certainty quake.

And when you came to write about her, you realised more bitterly than ever what an insincere form the Victorian novel was.

You knew where that word 'insincere' came from — the honest sculptor did not use wax to hide the blemishes in his work.

And the Victorian novel was held together with wax—the wax of the regular climax to suit the serial form and the wax of the grotesque turning points.

The Tess of this conventional fiction yields to Alec and later murders him— the real Tess is found elsewhere in the book.

You see her driving on that fateful morning with the load of beehives and digging the garden until it is almost night and happy in the dairyfarm where she meets Angel Clare.

The Tess in the waxen parts is totally out of character—she is an intruder from the world of 'Murder In The Red Barn'.

You, Thomas Hardy, cannot be blamed for that: when you are dealing with publishers, total integrity is a passport to silence.

Nor can you be blamed for that wooden hero you christened Gabriel Oak.

Nor can you be blamed for the web of coincidence woven around the characters in 'Far From The Madding Crowd' — the Victorians demanded plots even stranger than the truth of their own age.

The most authentic portrait, slight though it is, in the book is that of Joseph Poorgrass.

He is more than a member of the alehouse chorus—he is a recognisable member of the invisible army of the fields.

Fate has thrown the dice for him so that his days will be given to long labour—and his intermissions will be mainly sweet only because that labour has temporarily ceased.

You knew him, Thomas Hardy: he was your cousin—and though you never lifted a hand to help Joseph Arch in his great work for the farm labourers, you understood their world very well.

You symbolised it in what Joseph Poorgrass looked on as the greatest event in his life.

It came at the circus when after Dick Turpin's Black Bess had 'died' on the stage, he was one of the men from the audience who helped to carry her out.

And all the time, while you were working at your novels, you were distilling your experiences into poetry—that was your uncompromised truth.

The novels are not to be devalued because of their enforced artifices: they are all good—and one of them, 'Jude The Obscure', would be great but for its relentless pessimism.

That was a book you could afford to write for yourself—you were then powerful enough to be able to ignore both publisher and public.

And in all the other books there is greatness despite the constrictions — but yet, Thomas Hardy, when you looked back from the plateau of your ninth decade, it was the poetry that gave you the deepest satisfaction.

And you were an odd man — occasionally you said to yourself: "I will show the literary world that I can write their kind of poetry too."

And so you wrote pieces that came out of 'The Golden Treasury' — and indeed went into it when it acquired a supplement.

'Weathers' is such a piece—and it is marvellous of its kind.

But your deeper utterings are more ruggedly crafted — they are the work of one who was the son of parents whose home language was not Victoria's English but the ancient speech of Wessex.

And even though you built yourself a big house and moved in high society, you were never far from the life of your cousins who worked in the fields.

You remained forever fascinated by the way that the simplest lives can be touched by greatness.

For some of your neighbours it was being snatched away to fight in a foreign land — the Napoleonic and Crimean Wars brought many a Wessex peasant to a world far removed from his well-trodden fields.

For others the battles were less-obviously epic — as you said: "At the graveside of even the humblest man you see his life as dramatic."

And though T.S. Eliot dismissed you in that purblind essay and deemed you a nay-sayer, you were a mighty lover of life. In your crabbed way, you were an affirmer.

There is a little poem you wrote about the unmarried girl whose child dies — your philosophy is in it.

She was ashamed of him while he was alive; now that he is dead she sees how wrong that shame was.

And there is that marvellous poem about the felling of a tree—it is as powerful an intimation of death as is George Orwell's 'A Hanging'.

And you felt that your genius had not got its just reward—there is a lot of you in Jude Fawley.

You were vain and mean and ruthless — and anyone who thinks that great art is always born out of sweetness knows nothing about you, Thomas Hardy.

Evening Press, Tuesday, May 10th, 1977

Hardy — The Chronicler of Quiet Fields

Perhaps the most likely explanation of Thomas Hardy's new popularity is contained in that familiar saying about the difficulty of keeping good men down.

Yet it is hardly so simple: that concept implies a world free from the wilful agent we call chance. And the filming of 'Far From The Madding Crowd' was possibly the key that re-opened the great treasury.

Yet that also is perhaps too simple an explanation: the decision to bring Gabriel and Bathsheba and the rest to life on the screen was hardly unlinked to a shrewd awareness of mental climate.

D.H. Lawrence has come back too, although a generation ago he was sentenced to obscurity by Cyril Connolly, just as Hardy had been by T.S. Eliot.

And the bond between Hardy and Lawrence is perhaps the best clue to their return: in this age when artists seem obsessed with outsmarting one another, it is a benison to discover that there are those who know the difference between being clever and being intelligent.

Most artists are perhaps aware of that difference but not all have the courage to cross the divide. Hardy and Lawrence had it—and that is the salt that keeps their work as alive as when if first came from the presses.

The influence of Hardy on Lawrence is palpable; it is almost impossible to find any literary influences at all in the work of Hardy himself.

If there are influences, they are more likely to be from painting: Hardy started out as a painter— and it is not fanciful to trace the mark of Turner and Constable in his writing.

And if Hardy seems to belong to no literary tradition, it may be due to his conceiving in visual images rather than in words.

And painting may have influenced him in another way too: Millet and Courbet, possibly for the first time in European art, had treated the invisible rural poor as beings with their own peculiar depth and dignity.

Another factor may have caused him to be a rather reluctant dweller in the literary tradition: the queen's English was not altogether his native tongue.

Hardy had an unusual childhood in that he spent almost as much time in a neighbouring big house as in his own—and though his mother was very much of working class stock, she discouraged the use of the local dialect, yet his deep familiarity with it argues frequent early contact.

And the common speech of Wessex was as far removed from the standard language as were the daily lives of those who spoke it from the lives of the queen and her court. It survives in literary form in the poems of William Barnes.

Hardy wrote a few early verse-pieces in it and was tempted to make it his literary language. Sensibly he changed his mind and thus got what never harmed any artist—the stimulus of a wide public.

And yet there are times when one feels that he was never fully at home in the standard language—or perhaps it is more accurate to say that he had a less than perfect feeling for words.

There are few memorable phrases in Hardy: he could not create a picture in a few strokes. There is nothing to compare with Lawrence's 'hedges showing the birds' nests no longer worth the hiding.'

And again we must qualify that judgement: it is true of his prose but not of his poetry.

And despite his precise painter's eye, one is most conscious of his awkwardness of language when he is attempting to capture the essence of a place.

Georges Simenon, surely the most under-rated writer practising today, can summon up a town or a street or a cafe in a few lines: Hardy fumbles stubbornly on and sometimes seems to intensify his effort as his desperation increases.

Nor did he ever compose a novel with the unity of 'The Great Gatsby' or 'As I Lay Dying' or 'The Sun Also Rises'—but that was in part due to the vogue of the serial in his day.

Nor are his plots particularly convincing: 'The Mayor of Casterbridge' is by no means his least probable story—yet a summary of it reads like a send-up of the Victorian sensational novel.

Again the fault was not altogether his own: he grew up in an age when life seemed bent on outstripping melodrama. The Tichborne case could hardly happen now — nor is it likely that there will ever again be a legal showpiece to compare with the Camden Town murder trial.

And if people in Hardy's novels have a habit of disappearing and being believed dead and then turning up at decidedly awkward moments, it should be remembered that in Sussex in the eighties a man returned home from Australia and caused no little embarrassment to the judge and jury who had sent a man to the gallows for murdering him three years before.

And yet even allowing for his bilingual childhood and the bizarre age he lived in, it must be admitted that Hardy is an awkward writer and a clumsy plot-maker. He is also a great novelist.

And he would have been greater but for the conventions that shackled the novel in his day: he was a professional writer and had to pay homage to the story that hinged on dramatic turning points.

And yet at least once he escaped from that tyranny: within the conventional framework of 'Tess of the D'Ubervilles', a story that Cecil B. de Mille might have deemed too improbable, there is a kind of stowaway book in which truth is uninhibited by the demands of the publisher.

There are two Tesses: one, she who is seduced and who eventually kills her seducer, belongs to the world of 'Murder In The Red Barn'; the other is a real person and any resemblance between the two is purely coincidental.

When one has dismissed the nonsense (the flight of Tess and Angel might have been scripted for a Richard Burton and Elizabeth Taylor spectacular) a marvellously real and moving picture of a rural working girl's life remains.

This real Tess is Hardy's greatest creation: no doubt she is to some extent based on more than one person, but the main source was his own mother.

And the tragedy is not in Tess being hanged: it is in a sensitive girl having to endure a life not far removed from slavery.

In most of Hardy's novels there is this sense of the real world underneath the melodrama — and what emerges most powerfully is the Wessex countryside. One thinks of what was said of Lawrence: "He could describe people in a field — and when you had forgotten the people, you remembered the field."

What makes this world of Wessex so memorable is the acuteness of the observation: one is reminded of Constable's famous 'Cottage in a Cornfield' where the part of the field in the shade of the house is less ripe than the rest.

Hardy's effects tend to be cumulative, but there are some tremendous single passages. The greatest perhaps is that describing the driving in of the cows by Tess in the 'non-human' hours of a Summer's morning.

There you find Hardy's extraordinary sensitivity to the natural world. There are other passages hardly less memorable, especially that in 'The Return of the Native' where he evokes the night of the bonfires on Egdon Heath.

Some of these great passages have little apparent relevance—and one feels that for Hardy they were rewards he gave himself to compensate for the sheer plod that went into the novels. They were, if you like, indulgences in a kind of poetry.

And eventually the day came when, in seaman's idiom, he took a shore job: 'Jude The Obscure' was to be his last novel — from then on the storms that the later works had aroused did not unduly bother him. He was financially secure; he wrote to please himself.

It is hardly true to say that he became a poet in his fifties: he had always been a poet — now he poured his mental energies into verse.

It was not only his security that caused this change: the death of his first wife had an effect on him that seems surprising after so ill-starred a marriage. Perhaps it was his awareness of their tragedy that affected him: his poetry is like an illustration of Virgil's 'lacrimae rerum', of the world's sadness.

And if the language of his poetry retains some of the old awkwardness of his prose, it is yet purer, more lucid, more economic—it is as if he had found his truest voice in a form less inhibited by convention and the publisher's reader.

And poetry gave play to his very fine feeling for rhythm. This was part of his cultural inheritance: Dorset was great dancing country and Hardy was a country fiddler.

And when all is said and done, the essential difference between prose and poetry is that in one you can mark the rhythms—and this gave Hardy a dimension that lightened and distilled his language.

You will find occasional seeming awkwardness in his poetry but it is deliberate, sprung from his belief that effectiveness can sometimes be achieved by little disruptions of a regular pattern.

And it is one of life's little ironies that what possibly seemed to him his minor career produced the better fruit: Hardy is a great but deeply-flawed novelist—he is a superb poet.

One comes on some of what he may have thought the most profound passages in his novels with embarrassment; the simplest of the poetry is the most profound.

And he was at his best when obeying his own dictum about the significance of the humble, the things that "will go onward the same though dynasties pass."

———————

Evening Press, Tuesday, May 24th, 1977

The Quest for a Key to New Zealand

In Italy during the last war there was a partisan who had been cut off from his comrades and who carried on the struggle by stealing down from the mountains in the night and on any convenient surface chalking 'Non, Non, Non'.

In Wellington one time there was a man who was akin to him in his own peculiar way.

He was old and little and poor but this did not prevent him from waging war on what he considered the enemy — in this case, the New Zealand mental climate.

His main tactic was to storm into a bar where the better-off among his compatriots were drinking and shout 'John Keats is the greatest poet that ever lived.'

It was a subtle charge: the dominant attitude was not favourable to poetry — and even if it had been, Keats would be most unlikely to become one of New Zealand's favourite poets.

You can read about this strange partisan in 'Once Is Enough', the tortuous and yet exhilarating self-portrait of a great Antipodean writer as a young man.

Frank Sargeson is not well-known in this part of the world; of all New Zealand's writers only Katharine Mansfield is — and she was very much part of the London literary scene.

His lack of European fame may be due to more than his physical remoteness: perhaps he is too honest, too roughedged, too unglossily packaged.

And perhaps he is too unlike our idea of a New Zealander: we tend to have neat images of other peoples — and best sellers are often works that pander to these.

Thus to many 'Robbery Under Arms' is typical of Australia — but 'The Overlanders,' far less dramatic, far less 'romantic,' is a much truer picture.

And when you first meet Frank Sargeson's work, you experience a sharp disappointment: you expect a great sense of freshness from this new world under the Southern Cross — and you find it is not very much different from the old.

The fault is not his but yours: of course it is a very different world — but not in the way your over-romantic imagination pictured it.

The stars are different and the scape of the land and the way the birds sing and the rain falls—but the dominant people are European and brought the accretions of ages with them.

In Frank Sargeson's youth the bush was being slowly dominated and a new land created — but if the earth under his feet was raw, the ideas that surrounded him were stale.

The dominant moral code was a kind of tired Puritanism. Its exponents preached that the country faced two great evils — the brewers and the Catholics. Under the shadow of that terrible duo young Sargeson struggled towards manhood.

His father was a genuine Puritan: he worked long beyond his office hours as a town clerk and believed that books and plays and films and a lot more were of the devil — drink, of course, was so monstrously evil that it could hardly be mentioned.

His mother he depicts as even worse: she was a sham Puritan, a life-lover whom marriage had straitened and whose code of behaviour was all the more negative because it had no moral centre.

All this is enough to disillusion anyone who thinks of early New Zealand as a land where life might start anew.

But there is more disillusion to come. Young Sargeson attempted to find himself and his country by joining an uncle who had dedicated himself to farming.

Sentimentalists who rhapsodise about 'the good earth' should read 'Once Is Enough': the earth is an unrelenting guerrilla that needs only a little truce to re-assert its primal nature.

Young Sargeson's uncle was a very skilful and intelligent husbandman — but even he at times felt that the battle was unequal.

You laboured mightily to clear away what is called the bush; you built your roads and bridges and houses — and then congratulated yourself that the hardest part was over.

But then came the rains—and suddenly the creeks that had hitherto borne the excess of water safely to sea now seemed to rebel and washed away your roads and bridges and houses.

What was happening was that there was no longer enough vegetation to hold the rainwater and release it gradually — it ran off the bare land and the creeks overflowed.

That unforeseen hazard was eventually overcome, but the battle was far from ended.

No sooner had one species of unwanted vegetation been extirpated but more, some perhaps dormant for ages, sprang up to take its place—nature seemed not so much neutral as hostile.

And eventually even these were beaten or at least kept at bay — and then the sheep were ambushed by diseases unknown in Europe. New Zealand was not exactly an unblemished promised land.

And there were yet other enemies. A man might sweat all day and every day for a very long time and feel that in the end he had achieved a solid victory.

But it was not as simple as that: his land might be mortgaged — and back in London might be men in pinstriped suits who by a little adjustment in the interest rate could wipe out his supposed gains in one fell stroke.

When Frank Sargeson first joined his uncle, he did not see on all sides a new world unfolding: most of the neighbouring farms had been abandoned and were reverting to the bush.

Those who stayed on were mostly people who had been tenant farmers in Britain and were ambitious to acquire their own land.

Frank Sargeson's uncle remained too — and it came as a deep blow what was the force that drove him on to live such an arduous and lonely life.

You will find a hint at the same paradox in 'Sunday Too Far Away' — that sometimes the great pioneers, the men who perform the epic labours in outback and jungle and desert and bush, are forced into heroism by doubts about their sexual proclivities.

And so young Frank found no simple answers 'up country.' He went on with his studies and became a lawyer, though vaguely aware that it was not his vocation.

Determined stints of travel on the European continent and reading in the British Museum did not bring any instant revelations.

One is reminded of Thomas Wolfe saying that when away from America, he was the more aware of it — and 'hungered for the feel of the peeling handrail of the board walk in Atlantic City.'

Frank Sargeson, back in New Zealand, drifted along, working at his law job, watching or playing rugby or cricket or tennis.

And then one evening the little old man burst into the hotel bar proclaiming that Keats was the greatest poet that had ever lived.

Frank Sargeson attempted to speak to him but the old man pushed him away as if saying: 'Life is no place for dilettantes.'

It was a kind of Damascus — sudden and total conversions are rare outside mythology. But the young man became more aware of his own mental and spiritual passiveness.

His struggle back towards a vision of a shaped and purposeful life was assisted by another rather untypical New Zealander.

He was a man who worked in the small hours tidying up a factory — at night he read in the public library. There Frank Sargeson came to know him.

And the story of their relationship is the best part of a maddeningly uneven book.

This new friend was the kind of person most of us have known—an unremitting questioner who had built up his own mind by Charles Atlas-like endeavour.

Frank Sargeson was now bent on being a writer—and the irony was that his new friend, devastatingly honest and powerful of mind, felt that he himself could not write.

In his pocket he kept cuttings of letters of his that had been published in newspapers — but at home he had more never finished because he had despaired of saying exactly what he meant.

And so Frank Sargeson had to set out alone to clear away the bush and make his own literary earth.

For a start he returned to spend a long spell on his uncle's farm. Now, he was more clear-eyed, more conscious that someone, however crudely, should make a start at 'finding' and expressing New Zealand.

And when you read his work, especially 'Collected Stories,' you will probably agree that he hasn't done too badly at all.

———————

Wordsworth . . . simple giant

"God," wrote William Cowper, "made the country—and man made the town." And he uttered that naive concept long before the brilliant inventions of the eighteenth century had been used to enslave rather that to liberate.

God did not make the country: man took its primal shape and over a vast span of time he hewed and dredged and ditched and drained and domesticated it. And the town was not inevitably morally inferior because man-made. One of William Wordsworth's best-known poems illustrates that truth.

Is there a better evocation of London than 'Upon Westminster Bridge'?

The city now doth, like a garment, wear
The beauty of the morning: silent, bare,
Ships, towers, domes, theatres, and temples lie
Open unto the fields and to the sky,
All bright and glittering in the smokeless air.

In the popular mind Wordsworth is a "Nature" poet — and yet he could say of London "Earth has not anything to show more fair."

It is true that the London he knew was not the huge and ragged-edged conurbation of today but a compact city by the river.

But it was no urban paradise: a big part of it was dockland — and even its most fashionable quarters were less than salubrious.

Yet it inspired Wordsworth to create a noble poem — because despite all its faults, it was a true city: it had an intimate unity — and an abundance of that beauty so necessary for man's spiritual well-being.

Wordsworth would have understood Sherwood Anderson who a century later said: "How can a man love his country when he does not love his own house or the street on which it stands?"

Anderson was talking about the industrial slums of the United States — "as lovely a land as ever lay out of doors—on which we have put the stamp of ourselves for keeps."

And if Wordsworth is a "nature" poet, it is because in his years of greatest ferment the Industrial Revolution was beginning to spawn its horrors.

"Nature" is a much-misunderstood word: Oscar Wilde was right when he said that you can be as close to nature in a café as in a jungle.

And there is not the slightest element of mysticism in Wordsworth's oft-debated: "Nature never did betray the heart that loved her."

He is talking about absolute truths — and the need to abide by them.

That, too, was the creed of the American Indians — before "civilisation" descended upon them.

And those among the Indians who have retained their spiritual wholeness believe that the tragedy of White America is its alienation from these truths.

It hardly needs saying that White America has not co-operated with the earth — it has exploited it.

Man makes the country—and the images he creates inevitably affect him.

And when he both despoils the country and builds anti-human cities, he is gnawing at his umbilical cord with nature.

R.S. Thomas, in his introduction to 'A Choice of Wordsworth's Verse', writes:

"As he reclined in a grove somewhere in 1798, it grieved his heart to think 'what man has made of man. To many in these islands nearly two hundred years later it may be grievous to think what man has made of nature."

And Wordsworth would grieve even more to think what man has made of man in the years between — certainly the profound faith of the eighteenth century in moral evolution has not been fulfilled.

Who was William Wordsworth, then, that he took it on himself to ask such big questions and to attempt such total answers?

He grew up far from the intellectual centres of Oxford and Cambridge — he was that formidable being: - a provincial with an infinite capacity for being underawed.

Patrick Kavanagh had it. So had Emile Zola: Wordsworth evoked London in a sonnet — Zola attempted to storm the spirit of Paris in a series of novels.

And in Wordsworth's young manhood the mental tide was running against the intellectual establishments: the taking of the Bastille had been a ragged piece of black comedy — but it was an enormously potent symbol.

How much the French Revolution affected Wordsworth has been endlessly debated — Herbert Read, a paragon of honesty, saw his brief involvement with Paris as profoundly influential.

It is doubtful. It is more likely that his mind was already moulded into its general shape — that it was like a rocky field which the flood waters touched but did not permanently change.

And it is doubtful if the revolution influenced his theory of poetic language — it was probably more a turning away from the excesses of those who too assiduously imitated Milton and Pope.

Wordsworth, like most formidable provincials, was a loner; you could never imagine him part of any school of poetry, no matter how loosely the term is used.

The intellectual habits acquired in young manhood rarely change — and Wordsworth probably owed more to a fairly solitary boyhood in the extreme North of England than to Paris of the ferment and the barricades.

When all is said and done, his values are the peasant values — he expresses memorably what the peasant deems hardly in need of expressing at all.

Perhaps it is better to say that the peasant expresses it—but not in words; his work is sufficient a statement.

Wordsworth clearly wished to be a popular poet — one whose verses would be part of the treasury of shepherds and ploughmen and wood cutters and all those whom he saw as the common people.

He hardly achieved that ambition; it was almost impossible by then for a serious poet to be popular in a wide sense — and he wrote nothing that achieved the folk-status of "Elegy in a Country Churchyard".

His declared intention of writing in the language of the common people was obviously part of that ambition — it was a misguided notion and was quietly abandoned.

Those who speak about folk poetry and folk music are equally misguided — there are no such things.

The concept of a folk mind producing poems and songs and airs is very attractive—but hardly bears examination.

The gifted individual creates — and in the course of time other gifted individuals may make little improvements in his work: It is "folk" only in the sense that it is accepted.

And so it is with language: that too is created by individuals. And it is nonsense to say — as is often said — that Milton did violence to the language. He used it to articulate an uncommon vision.

Some of Wordsworth's poems seem perfect illustrations of his own theory about poetic language.

It is the first mild day of March:
Each minute sweeter than before,
The redbreast sings from the tall larch
That stands beside our door.
There is a blessing in the air,
Which seems a sense of joy to yield
To the bare trees, and mountains bare,
And grass in the green field.

But that little poem, "To My Sister", is only a simple description of the day and his plans for it. It is a long way from the glory of "The Immortality Ode."

Hence in a season of calm weather
Though inland far we be
Our souls have sight of that immortal sea
Which brought us hither
Can in a moment travel thither
And see the children sport upon the shore
And hear the mighty waters rolling evermore.

There is hardly a word in this passage that might not be used in common speech — but nevertheless it is in the language of high poetry.

William Cowper came much nearer to realising Wordsworth's ideal of poetic language — partly because he was greatly influenced by the Methodist hymns, the popular poetry of his day.

There are passages in "The Task" that Ernest Hemingway might have envied — especially the famous description of the woodcutter and his dog going out into the snow.

And John Clare achieved astonishing effects with the plainest of language.

But Wordsworth was seeking something too big — the language of description is not always sufficient as the language of concept.

His theory, however, had one great benefit — it served as a corrective.

Shelley would have benefited from it: too often he forgot that he had an audience other than himself.

Often we hear it said that he had mighty ideas but could not express them—but there are no ideas until they find words.

The art of poetry lies in finding expression for the mental impulse — and Wordsworth often almost convinces you that Coleridge's curious definition of poetry is sound: "The best words in the best order."

But then you realise that the words and the order seem so good because the whole is infused with such passion and sincerity — and that Tennyson is much closer to that definition.

Wordsworth was a patient craftsman — but he was no embellisher: he was like a builder of plain stone walls.

The beauty of the walls is in their strength — a poem, like a wall, has its own truth.

And truth was the object of Wordsworth's great quest.

What did he bring with him after his journey through the forest? He came out with his early intuitions reinforced.

The progress or regress he witnessed, as the Industrial Revolution grew, made him more convinced that capitalism and the city were hostile to man's real needs.

He believed more and more in contact with the elemental forces.

D.H. Lawrence in his role as prophet was to preach that doctrine — but in a hysterical form, and in modern America it is the seminal idea of what is called the Green Revolution.

Wordsworth's doctrine is most concisely expressed in the familiar sonnet "The World Is Too Much With Us" — and more expansively and obliquely in 'Michael.'

That poem is like a long short story — and has an ending not unlike those associated with Guy de Maupassant.

It tells of a shepherd and his wife whose hope of the continuity so precious to them was invested in an only son.

The father plans that he and the boy will together build a new sheepfold — "a covenant 'twill be between us".

They start the work and make good progress — but before it is completed; the young man goes off to seek his fortune and promises to return.

For a while he gets on well — but then falls among bad companions and eventually has to flee the country.

The news of his disgrace reaches his parents. The sheepfold remains unfinished.

'Tis not forgotten yet
The pity which was then in every heart.
For the old man — and 'tis believed by all,
That many and many a day he thither went,
And never lifted up a single stone.

Evening Press, Tuesday, June 5th, 1984

I too have lived in Arcadia

Most of us began in the same way: we used to catch the sprats with muslin nets and put them into jam jars.

They weren't really jam jars: they were big glass containers that had held boiled sweets; throw in a fist of gravel and sand and a few stones—and you had a little aquarium.

The nest step in our education was the acquisition of a wire net. You didn't buy it—in those days you bought nothing that you could make yourself.

Of course you had to buy the wire—the kind of wire used to keep chickens cooped up until they were fit to go out into the big world.

'Lattice wire', I believe, it is called: we called it 'lettuce wire' but that hardly mattered: shaped into something like a giant's purse it was very effective.

And then like good old Willie Yeats you went to the wood — not to a hazel wood but somewhere that you cold get a long pole of ash.

Ash was the best — just as it is the best for making a hurley or a swingle tree. It is the best because of all trees it is the most alive.

And so you fitted the net onto the pole and you went down to the river. The job needed two — one for the net and one to drive the fish towards it.

The latter was known as the poker: he had a hard job and could end up the day with blistered hands.

He had a long pole, pointed at the end. With it he poked the nooks and crannies where trout live when they aren't out feeding in the stream.

He was like the hod carrier: the man with the net was the artist — at least in his own mind.

Because the poking stirred up the gravel and the sand and the mud, the man with the net couldn't see the fish; he had to depend on 'feel'.

The 'feel' came to him from the pole — and he had to know a stone from a fish; otherwise he might pull up the net too soon.

It was an enthralling game and you felt that you had reached manhood though you might be only an immature seven.

It was the best of all in a flood. A few of us who were convinced that we were in the super league used to say that any fool could catch fish in low water: the flood was the big test.

None of us experts — we were a co-operative numbering four — ever wore wellingtons when working at our craft: in our own crazy way we sensed that they were dangerous.

They were slippery. And if you fell in the river, they could fill up and take you to a watery grave.

And so we would set out in our bare feet. At the river you rolled up your short pants as far as you could. Then you waded into the flood — and you were in business.

When the flood comes, the fish leave their nooks and crannies: they have no choice — the force of the water makes their homes untenable.

They seek quiet little bays, places where normally there is no water at all. Cows' drinking places are ideal homes in a flood.

And in these conditions the poker becomes redundant: you don't poke — you put in the net and scoop.

And many's the time we saw such a dip-and-lift produce a miraculous draught of fishes.

It isn't only reporters who get scoops.

'Netting', as we used to call it, is the second step in your apprenticeship. You never really tire of it but you know that fishing with the rod is infinitely superior.

And so some day you go to the wood again and bring home a nice pliant pole and you trim it into shape.

And then you attach a few very small staples at regular intervals. And then you get a spool that once held thread — that became a reel.

A piece of wire served as a handle. It was no bother to make hooks. You had to buy the line and the gut — and then you were in business.

I am certain that every angler remembers clearly the moment he caught his first fish.

There was a man in our town who got a place in the local folklore on account of his reaction to this momentous experience.

He was a grown man when he took to the river. On his second day he caught a small trout — and ran about a mile home to show it to his mother.

And once that passion takes hold of you, you will never get rid of it. I have known men who had to retire from angling — but the love remained, growing stronger by the day.

And if I ever envied anyone, it was Izaak Walton: his life — at least as we see it — was beautiful.

His shop in London gave him a good living and a fair amount of spare time.

He loved angling and poetry. He wrote a classic about his avocation — and he wrote the biography of his friend, John Donne.

I know that Walton lived in turbulent times and that London was then a violent and pestilential town.

And he took 'the wrong side' in the civil war.

And yet he was lucky; he lived in an age when rural waters were pure — 'modern' or 'scientific' farming was centuries away.

To devote your leisure time to thinking and experimenting about something you love isn't a bad way to spend your days in this world.

Another Izaak — the mighty Newton — devoted his life to physics. Thus he may seem more important that the man who compiled 'The Compleat Angler' — that would be a superficial judgement.

Why is angling such a passion?

I suppose part of the answer is that it is a surrogate for a primal activity — in your own mind you are man the hunter.

There is more to it: water is an endlessly fascinating element — and angling is truly, a going into the unknown.

The humblest river abounds in mystery—and there is always more to a river than even experienced anglers suspect.

I will try to explain. Let us suppose that you know a river as well as a human can; you know all its pools and runs and bays and the size of the fish they usually hold.

You flatter yourself that you know the river — but some day you could get a surprise: a drought might come and the river's bones be exposed — and you might see fish bigger than ever you had suspected.

You can have a similar experience when a small river is being dredged: the bucket may bring up trout bigger than ever you had seen there.

I am tempted to make a comparison with the process we call life — and it wouldn't be pretentious.

I am also tempted to suggest that anglers are a special breed: almost all of those I have known were decent people.

There is a touch of the poet about them; they have imagination — and they have aficion.

'Aficion' brings us to Ernest Hemingway: it was a great word of his.

And it brings us to his famous statement: "Enthusiasm isn't enough."

He was talking about sea-angling but the dictum fits a multitude of pursuits.

Of course enthusiasm isn't enough: aficion signifies more than passion; it implies the knowledge that passionate interest brings.

And the best anglers are those who listen. The best one I ever knew would listen to a child fishing with a bent pin.

There is a moral there that applies to the so-called real world — but beyond indicating the danger implicit in fixed ideas we will not pursue it today.

Instead let us cast away solemnity and come with me to my favourite river.

It goes by several names — the Gleannsharoon river is my favourite name for it.

It is a modest enough stream in a dry summer but I once saw it sweep away a big farm-cart that had been left too close to it before the flood.

It flows from the hills north-east of Castle Island; the dredgers haven't yet got beyond its lower reaches; it is a 'natural' river.

It is a mixture of spring water and bog water — and so excellent for trout.

Its bed is generally sandy — and in its lower reaches you will find flounders. They are known locally as 'fluke'.

It is usually at its best on the day after a flood when the water is well up but clearing: 'beery water' is the local term.

And then with a small lob-worm on a fly-hook you can have a good day — or you might catch nothing.

It depends on the mood of the trout — if they are hungry, you will do well.

If they are sated — as they sometimes are after a flood — all your skill and guile and changing of bait may prove in vain.

Our stream is a tributary of The Maine, the big river that meets the Atlantic about fifteen miles away.

And their congruence is a great place to meet fish in a rising flood.

There is a little bay there, dug out to make it easy for cattle to drink — it is ideal in high water.

My usual practice when fishing in a flood was to start up near the hills and work along down. Eventually I would arrive at the meeting of the waters — and become part of a standing army.

And there we would exchange our experiences and enter into a kind of friendly competition.

There were days when you needed infinite patience — and days when the water was boiling with fish — and enthusiasm for once was enough.

And there were the quiet days when the water was low. Then the fly fishermen and the upstream-worm men came into their own.

These were the days when — as a great friend of mine used to say — you should know what the fish are thinking.

His name was Danny Horan — and some day I will write him a poem "as passionate as the dawn."

Evening Press, Tuesday, November 18th, 1986

Paris — the promised city

Paris in the spring — aye, it's a lovely image — but not too long ago I was in the capital of France on a May day and the rain was coming down in torments. And it was cold too.

And by the night's end — which means some time in the morning — I felt like a wolf that had just been dragged out of The Seine.

And yet I was outrageously happy. I spent the last few hours in a little café, solving the world's ills with a few new-found friends. And the wine was good.

I couldn't tell you its name or its year or even its century: it came from unlabelled bottles and was none the worse for that.

All I know is that it was dark-red and tasted well and, as William Cowper said about tea, it cheered rather than inebriated.

Anyhow, I felt that I had come unto The Promised Land or rather The Promised City: Paris affects me thus.

To be there is like a little truce from life's battles: a friend of mine has described this feeling as General Paralysis Of The Seine.

But sometimes I wonder how would it be if you had to work there.

And I suspect that my vision of Paris is wildly romantic and probably justifies him who said that imagination is as good as a physic to a fool.

In my less ecstatic moments I know well that Paris is a place where the mass of the people work hard for modest pay and where many live in cramped conditions in blocks of flats almost as dreary as those in Moscow.

You may remember the old man in Emile Zola's 'L'Assommoir' who 'lived' in the cubbyhole under the stairs in a tenement.

He entered the story only because a party in the house found that they were thirteen at table — he was brought to allay superstition.

Zola drew Paris from life. In his day — and for long after — you could see barefoot women, not all of them young, drawing bread-carts: they could go up hills so steep that they were dangerous for horses.

And if the Parisians seem to have an obsession with food, it is not because they are all gourmets or gourmands or both: history has taught them the true meaning of 'fruits of the earth.'

But 'living frugally' doesn't mean that you are on iron rations: it means that you are making the best use of the available food — or at least it did.

In the older sense of the words, the French live frugally — and the people of Paris have especial cause: at least thrice in modern history the citizens of the capital saw more dinner-times than dinners.

They didn't starve during the German occupation in the nineteen-forties: neither had the great majority to worry unduly about their weight, unless it was going dramatically down.

During the siege that was the culmination of the Franco-Prussian war, the people of Paris boiled boots and shoes to make soup.

And in 1789 sheer hunger was almost certainly the immediate cause of the event that is generally deemed the start of the revolution which in turn is generally deemed the seed-bed of modern Europe.

Acres of words have been written about the causes of the French Revolution: most analysts believe that it was inevitable — but chance may have played a part.

Hailstone is usually experienced in the interim when Spring is ended but Summer hasn't quite begun: the harvest of 1788 proved an exception; hailstones devastated the cornfields in the plains of Northern France.

Bread became scarce in Paris. For its 'little people' it was their staple diet, accompanied by the kind of wine that comes in unlabelled bottles.

There were riots. They were bloodily suppressed. Ring-leaders were hanged in public.

A sullen winter followed; but for the religious orders, and especially The Vincentians, the very poor would have died of hunger.

The order brought and stored food from the more fortunate parts of France; the consequences were to prove savagely ironic.

It was obvious that the Mid-summer of the following year would be a crucial time: by then the fruits of the previous harvest would be running out — those of the new harvest would not be ready.

And so it was hardly a coincidence that the fourteenth of July became a huge time-mark in France's history.

Because of its symbolic nature the taking of The Bastille is reckoned to be the start of the revolution — but it had already begun.

The riots had started on the previous Sunday, the twelfth of July: the houses of the religious orders were attacked; the mob believed that they had stores of food; the belief was well founded but the interpretation was wrong.

Mobs are not noted for logic — and the great house, Saint Lazare, founded to practise the preaching of Vincent de Paul, suffered grievously.

The Father-General of the order described the invasion: "At 3. a.m. on Monday a band of furious plunderers, armed with guns and sabres and torches, forced our doors . . . from then until 5 p.m. the same evening multitudes of robbers succeeded one another.

"They were unafraid — Paris was without troops or defences of any kind. Everything we had has been destroyed . . . doors, windows, tables, beds, furniture, all smashed to bits.

"Any money we had plus the money belong to the students was taken. The library is no more. The refectory is a rubbish heap.

"The cellars are empty except for dregs of wine and the corpses of about a hundred unfortunates who drank themselves to death there."

And that grim passage probably gives us a fair idea of how the revolution began. 'Liberty, Equality, and Fraternity' was a stirring slogan — the reality was less glorious.

I suppose that most of us who were young at some time have cherished the occasional dream of being in Paris during the revolution — and especially in the early days.

The popular images are seductive: when not manning the barricades, you might be singing The Marseillaise in a café and exchanging amorous glances with some citoyenne of strong republican convictions.

But seemingly it wasn't that way at all: Paris during the revolution was an exceedingly dangerous place — and it wasn't only the enemies of the republic, real and alleged, whose lives were at risk.

There is the story of the seventeen-year-old Irish cabin-boy who went up to the capital for the craic and enjoyed himself hugely until he started to boast about the fighting qualities of his countrymen — he was hauled off to the guillotine.

In the years between the taking of The Bastille and Napoleon's taking of power, the history of Paris and France is like a huge skein of wool that was savaged by a horde of crazy kittens.

There was order — of a kind — otherwise the bread wouldn't have been baked or the vines tended or the wine put into bottles, labelled or otherwise.

And there were periods of comparative peace — but even in those interims the fear of the guillotine prevailed.

It was a time of denunciation: shopkeepers got rid of rivals; husbands got rid of wives; wives got rid of husbands; a few whispered words were sufficient to end a life, no matter how innocent.

There were trials but they were run on the lines favoured by Idi Amin: "Give him a fair hearing — but don't take too long about it."

That France staggered on and avoided social and financial ruin is a tribute not only to the French but to the endless resilience of life itself.

There was no lack of villains; some such as Jean Paul Marat and Maximilian Robespierre were brilliant but power crazy; others were merely corrupt.

And of course there were heroes: one was Monsignor Salamaun, The Pope's envoy in Paris; he went underground and slept in the Bois de Boulogne and saved many lives by tipping off people on the death list.

And the more you learn about those years, the more you ask yourself: "Was there a revolution at all or did France experience nothing more than massive and prolonged disorder?"

In comparison, the revolution that culminated in the execution of Charles the First was as well organised and as well carried out as a royal wedding.

Its leaders set out with a clear purpose — they achieved it: the 'divinity that doth hedge a king' was broken forever.

It is hard not to suspect that the real leaders in France weren't Georges-Jacques Danton or Camille Desmoulins or Honore Mirabeau or those who supplanted them but the mob of Paris.

Paris was the nerve centre — and its lowest denizens called the shots. Mirabeau sensed this very early in the 'revolution' but he was powerless to stop it.

And the old question arises. Is history dominated by individuals or by mass movements?

The general view is that the The French Revolution — such as it was — would have erupted irrespective of individuals.

It is argued that such was the enormity of injustice in 18th century France that an uprising of some kind was inevitable.

'Inevitable', however, is a suspect word in the context of human affairs: it smacks of determinism.

Earthquakes and avalanches and volcanic eruptions may qualify as inevitable — but hardly the course of human life.

Free will is a factor in the shaping of history; so is chance.

Marie Antoinette may never have said 'Let them eat cake' but she was capable of saying it: the 'little people' of Paris knew that.

If she had mingled with them and shopped in the markets, there might never have been an uprising — but it is likely that there would have been a constitutional revolution.

And if the harvest of 1788 in Northern France hadn't been devastated by freak weather, the chain of events that began with the riots of July 12, 1789, might never have started.

Napoleon is often produced as the arch-example of the individual stamping himself on history. But without the uprising would he ever have achieved eminence?

History's penchant for irony is savagely exemplified in the story of the little Corsican.

The rising that had 'Liberty, Equality, Fraternity' for its masthead ended with a dictator instead of a king.

When the French brought back Napoleon's body from St. Helena, they showed a marvellous magnanimity.

It is true that he restored order and unified France but then he almost destroyed it.

Some historians maintain that the nation hasn't recovered from the losses suffered in his wars.

So many young men who normally would have produced families perished that the population is still less than France could maintain — that is the argument.

Charles de Gaulle had this in mind when he preached that the country needed a population of a hundred million.

There are many ways of approaching history — Mary Purcell in a splendid essay in this year's Journal Of The Kerry Historical Society looks at The French Revolution as experienced by an individual.

Edward Ferris was a native of Ballymalis, a townland between Killarney and Killorglin, but spent most of his life in France.

He was a formidable scholar and a slightly mitigated scoundrel.

In turn he was royalist, republican, Bonapartist, and above all a survivor. He died a very wealthy man.

D.H. Lawrence: The Tiger's Spring

"He could describe people in a field," it was once said, "and when you had forgotten the people, you remembered the field."

At first sight this may seem one of those too-clever statements that are the bane of literary criticism—but, all in all, it is true: at capturing or evoking the physical world, D.H. Lawrence was a kind of magician.

He would not have objected to that last word—because the instinctive apprehension that made him so great is akin to magic. In one swift leap he could outstrip the plodding accumulator of facts: Lawrence was never "down pit" — and yet had a far surer grasp of the miner's life than had Zola.

Without that power to apprehend states of being other than your own, you cannot be a writer. That may seem an arrogant assertion—and yet its implications are generally taken for granted.

It has been said that no one has the right to pretend to lift the roofs off people's heads and see what is going on inside—but writers have always been doing it and there is little evidence of a movement towards reform.

Guy de Maupassant consciously set out to be "objective"—but happily was not; there is a modern French "school" that have even sterner aims—but they are about as convincing as a high jumper who wears shoes of lead.

Lawrence's great flaw was that he depended too much on what one might call inspiration; when he came to places where he had to leave the river and make a portage, he did so sulkily and carelessly.

For that reason most of his novels are ludicrously uneven: passages of bewitching power and beauty intermingle with slabs of writing so drab and slapdash that one suspects some eccentric publisher of having tampered with the text.

But the fault was his own — he would have no truck with those who might imply that genius entailed some capacity for taking pains. He loved to compare himself to the tiger; he loved the image of the swift spring — and forgot that animals of prey must watch and wait and stalk.

He would have agreed with Leopardi that great ideas are born out of heat rather than deliberate reasoning; he would have been less ready to agree with Croce who argued that until an idea is articulated, it remains a mental impulse.

Who was this man whom a generation ago Cyril Connolly dismissed as being of no permanent importance — but whose books appear in the shops with the regularity and freshness of Spring vegetables?

David Herbert Lawrence was born in a Nottinghamshire mining village in 1885. His father spent all his working life hewing coal; his mother's background was what might be called genteel. Their marriage was not notably successful—but probably happier than their most famous child believed it to be.

David was the delicate one of a robust family; he was often ill in childhood — and while his brothers learned the rough games of street and field, he became proficient at sewing and embroidery and painting.

At least one study of his life and work advances the theory that his genius sprang in part from suppressed homosexuality. If the man's essence is in his writing, that thesis is absurd.

There is a simpler explanation for his genius: the seed, of course, was born with him — and it was nourished by two major factors.

One was the place where he was born and grew up: it was a drab and cramped village — but it pulsed with the energy of the industrial revolution at its zenith.

And all around was countryside and that combination of raw power and quiet beauty can be extraordinarily heady.

The other factor was his delicate health: those who say that Keats would have been an even greater poet if he had not died so young, forget that it was his illness that made him so precociously great. Lawrence's marvellous apprehension of field and tree and mountain and river and sea and sky is that of someone whose senses are preternaturally sharpened by a feeling of time's rope wearing thin.

And it was in part a legacy of his background too: townsmen often have a deeper knowledge of the fields than their rural fellows — if only because they see them with a less familiar eye.

And in Lawrence's time the miners for several months of the year saw daylight only on Sunday — and perceived the rural scene like blind men whose sight had just been restored.

Another influence too greatly nourished the young Lawrence: the mental climate that surrounded him was 'chapel — something that in this country we hardly understand.

In its more extreme manifestations it is a bleak code — anybody who ever knew the joys of a 'dry' Welsh Sunday was not likely to cast a kind eye on evangelism.

But in its more sensible form it has an honesty and dignity and beauty that are compelling.

John Synge to some extent owed his love of language to the great treasury of 'chapel' hymns; it is not difficult to trace a similar influence in Lawrence.

And he was lucky, too, that the mental climate of his place and time did not inhibit friendship between boy and girl.

Lawrence in the Nottinghamshire of his adolescence — and that was about 80 years ago — knew the kind of freedom that is dawning in Ireland only now.

He grew up poor enough in terms of material wealth — but extraordinarily rich in the coalition of his own talents and the water that lapped around them.

He was a good scholar ('a great passer of examinations' as Aldous Huxley said about him) and got into Nottingham University and became a teacher.

But he did that only to provide a safety net in case he failed in his governing ambition — there may be no such being as a born writer, but from an early age D.H. Lawrence never doubted his vocation.

While he was still teaching, he wrote a novel, 'The White Peacock'. It is a kind of encyclopedia of Victorian piety and was written mainly to please his mother who was then in her last illness.

Two years later, in 1913, came the first eruption from the volcano that was the real Lawrence — and 'Sons And Lovers' for many people remains the most exciting novel in modern English.

Strangely enough in the light of Lawrence's enormous reputation you will not meet many American thesis poachers in the pubs of Nottingham or Ilkeston or Matlock — all great names in his heartland.

And when you read the first few pages of 'Sons And Lovers', you sense why: there are no Joycean puzzles to be unravelled — a world is unfolded before you with a directness that at first seems almost too simple.

But soon you know you are in the presence of greatness: you are reminded of what was said about Guy de Maupassant—that you forget you are reading and think you are watching life.

Some of Lawrence's later novels are so recklessly flung together that 'Sons And Lovers' by comparison seems a masterpiece of construction—but nowhere will you feel that the builder is putting numbered stones into place.

And yet it has a very real unity—it comes from the life of a family in a particular place and time.

The family are the Lawrences, as near as people in fiction can ever be to people in "real life"—and the principal character, Paul, is young David Herbert himself.

There is no formal plot but there is tremendous development. The watershed comes when the eldest son dies—and the mother's ambitions are transferred to Paul.

The son had been taken ill a few days before in London. His mother had hastened to see after him; the account of how the news of his death is broken to his father shows Lawrence at his finest.

The telegram has come and young Paul goes to fetch his father.

"It was a beautiful day. At Brinsley pit, the white steam melted slowly in the sunshine of soft blue sky; the wheels of the headstocks twinkled high up; the screen, shuttling its coal into the trucks, made a busy noise.

"Paul did not realise that William was dead; it was impossible with such a bustle going on."

Eventually his father comes up out of the pit.

The two walked off the pit-bank where men were watching curiously.

The father asks in a frightened voice, "E's niver gone, child?"

"Yes."

Evening Press, Tuesday, July 19th, 1977

D.H. Lawrence — The Eternal Quest

It is most unlikely that you will ever stop a hundred people in the street and ask them to nominate the greatest English novel.

But if you do, it is possible that 'Jude The Obscure' will be a front runner. It is easy to understand why — and yet I would not consider it even a contender.

And the reason is this: in the true sense of the word it is a morbid book— and though Thomas Hardy thought of it as a tragedy, it is not.

Tragedy implies failure to achieve some goal — but in 'Jude The Obscure' there is no real goal because the whole book is heavily shadowed by a feeling that life is essentially futile.

A book may depict the extremes of unhappiness and hardship and all the ills that man is heir to and yet not be morbid — if it is permeated with a sense of a greater world.

Liam O'Flaherty's 'Famine' is an example; so is Emile Zola's 'Germinal' — and that is one of the reasons why 'The Rainbow' is the most satisfactory of all D.H. Lawrence's books.

It was his third novel and was written when he was about 30 and is a splendid example of how happy the result can be when the fire of youth coalesces with the wisdom of maturity.

Lawrence had tip-toed into print with 'The White Peacock', a novel that remained becalmed by its desire to please; in 'Sons And Lovers' he had allowed his own uniqueness a kind of fumbling freedom; 'The Rainbow' is the bold articulation of one aware of his great powers.

73

We all tend to entertain a half-belief that somewhere in the world there are people more profound than ourselves — and those mythical spiritual fathers inhibit us. By the time Lawrence came to write 'The Rainbow' he was his own man.

One great lesson he had learned from Hardy: the first rule for a serious writer is that he must be prepared to make a fool of himself.

He must forget about the literary shop stewards; he must forget about fashion; he must give his own mental world honest expression.

And that is why, quite literally, both Hardy and Lawrence oscillate between the sublime and the ridiculous — and why Thomas Wolfe, despite the patient editing of Maxwell Perkins, is at once the best and the worst writer that ever came out of America.

'The Rainbow' is the most even of all Lawrence's novels: it has none of those arid passages with which he makes a careless causeway between one deeply-felt experience and the next.

It could not have because, above all else, it is an attempt to capture the overflowing life of a family who are rooted in the rich earth of that old and mature land where Derbyshire merges into Nottinghamshire.

It is a long meandering book but it has a powerful unity —it is the unity of a countryside marvellously evoked.

Lawrence's greatest gift was his ability to capture the essence of place: not only did he brilliantly depict the surface features — he caught the spirit too, what he would call 'the messages that arise from the depths of the earth'.

'The Rainbow' is suffused with this sense of the physical world — and you see its people as being influenced by it as surely as a fish takes its colour from the water in which it swims.

If the book has a central theme, it is the importance of self-realisation.

One is reminded of Sherwood Anderson's idea that most of us for most of our adult lives are smothered in other people's ideas and sentiments and attitudes — only in fleeting moments does the real person surface.

Lawrence sees the industrial revolution as deeply inimical to fulfilment. It stunts men and kills the 'true self'.

In the mine and the factory and the mill even the finest are unable to keep their sacred flame; the very necessity of making a living defeats them.

In contrast the farming family of 'The Rainbow' live as naturally as the trees in their own fields.

"Living on rich land, on their own land, near to a growing town, they had forgotten what it was to be in straitened circumstances.

"They came and went without fear of necessity, working hard because of the life that was in them , not for want of money.

"Neither were they thriftless. They were aware of the last halfpenny, and instinct made them not waste the peeling of an apple, for it would help to feed the cattle.

"But heaven and earth was teeming around them. And how should this cease?

"They felt the rush of the sap in Spring, they knew the wave which cannot halt but every year throws forward the seed ...

"They knew the intercourse between heaven and earth ... and the nakedness that comes under the wind in Autumn, showing the birds' nests no longer worth the hiding".

And yet all this being at one with the natural world does not guarantee fulfilment; some of the family feel the pull of a higher state — they are aware of men who though far less physically powerful yet seem as superior to them as they are to the bull and the horse.

And there you have the most constant of all Lawrence's themes.

His main characters, all of course to some extent based on himself, are all strivers.

And all seek to know modes of being other than their own: the clergyman's daughter fascinated by the gypsy is symbolic of the Lawrence quest.

His own life has been called a savage pilgrimage; it is as good a term as any to describe it.

He had to give up teaching on account of his health — and from the age of 26 lived as a writer.

But by 30 his best work was done: the novel that followed 'The Rainbow' marks the start of the decline.

'Women In Love' has a few good passages but it is an enormously silly book.

By then Lawrence had been adopted by a little coterie of idle rich who imagined that they loved literature.

The protagonists in 'Women In Love' go on endlessly about their souls — because they have nothing better to do.

Malcolm Muggeridge in the most ludicrous judgement ever uttered in the field of literary criticism — and that is saying something — dates Lawrence's maturity from this period.

With stunning perception he traces the origin of his new-found greatness to his wife, the daughter of a German nobleman.

If he had ever read her letters, he would have known that as a writer she was not even in the fourth division —she was strictly non-league.

Lawrence went on writing — and such was his genius that even his worst pieces are speckled with the purest gold. But after 30 the graph that had been ascending so beautifully began to waver and decline.

And the reason was alarmingly simple: he had cut himself away from his roots.

He wandered around the world searching for Utopia — he did not have to look beyond the heartland of his youth.

One thinks of Cavafy,: 'In those few fields or streets of your childhood there, no matter where you roam, you will live and die'.

Lawrence died when he was 44 — he left behind an enormous body of novels and short stories and essays and travel books and letters.

Even the feeblest things he wrote, even 'Lady Chatterley's Lover', a work that might have been commissioned by Mary Whitehouse, carry the stamp of his genius.

And the essence of that genius was his vision of man as being subtly and powerfully influenced by his surroundings.

He said it very simply himself: 'I am free — as free as a tree with its roots deep in the earth'.

Evening Press, Tuesday, September 27th, 1977

Edgar Lee Masters — Wry Diviner

It is something every artist dreams about — that some day he will discover the philosopher's stone and then his work will no longer be a part-derivative struggling but will shine out fresh and bold, honest and original, the articulation of what has never been expressed before.

And if he is a writer, the dream will embody an especial hope — that he will find out the truths of people's lives as certainly as if he could lift off the roofs of their heads and see what is going on inside.

It is a manic dream but without it no artist will ever amount to much: he will disobey Ezra Pound's commandment about not doing what has been done before.

Sherwood Anderson believed that most of us for most of our lives are not ourselves at all—that we are governed by accretions of accepted attitudes and concepts.

His abiding ambition as a writer was to find out the uniqueness that lies underneath the accretions, to divine the streams that run below the surface of the common day.

'Winesburg, Ohio' was the great adventure that came out of this ambition: it is a bold and honest attempt to capture the life of a small town in the Mid-West of America around the end of the last century.

And in form it seems marvellously original: it is a loose coalition of stories and what for want of a better word one must call pieces—they have a common setting in place and time and some of the people in them are met more than once.

'Winesburg, Ohio' was heroically original in its way of looking at life, but not altogether original in its form—it owed much to a work that is now half-forgotten, 'Spoon River Anthology.'

For Edgar Lee Masters this book was the product of the philosopher's stone: before its birth he was a successful lawyer but a seemingly failed writer—he was almost 50 and had eleven indifferent books behind him.

And that birth owed something to chance. Masters conceived 'Spoon River' when he was given a book by William Reedy, the editor of a magazine to which he was an industrious if not very distinguished contributor.

It was 'Epigrams from the Greek Anthology'—and Reedy gave it to him so that he could study its realism and verbal economy.

And Masters, who for years had been energetically going nowhere, now felt that he had a mould that would discipline his diffuse outpourings.

And slowly, fumblingly 'Spoon River' was born—and in its finished form was very different from the first conception.

Masters's 'epigrams' are in fact epitaphs—but on his verbal headstones and tomb-doors appear not the conventional pieties but the true life-stories of the departed.

'Spoon River' is made up of more than 200 pieces of free verse—every one a monologue spoken from the grave.

Its effect like that of 'Winesburg, Ohio' is cumulative: the fates of many of its characters are, as in Anderson's masterpiece, interlocking—and at the end you have an extraordinary feeling of having shared in the life of a community.

There is humour in 'Spoon River' but its hue is distinctly sombre: the very first poem or piece in the book is the epitaph of a robber and murderer who is buried next to a wealthy man who owed part of his success to adroit use of the bankruptcy laws.

Hod Putt says:

> ' ... I was tried and hanged.
> That was my way of going into bankruptcy,
> Now we who took the bankrupt law in our respective ways
> Sleep peacefully side by side.'

That wry little sermon begins the web that May Svenson describes as 'ingenious, multi-layered, dramatic, socio-historic, intuitive ... giving outlet to all his grudges, beliefs, indignations, insights, prophecies, discoveries of glaring injustice, revelations of life's mysteries and paradoxes—and his own eccentric philosophy.'

It is a fair enough summing-up, only that it gives an impression of turgidity—and 'Spoon River' is beautifully lucid and sure of purpose.

Its main theme is the repressiveness of small-town life. Masters grew up in Kansas at a time when the West had been won but when there was still a hysterical fear of the non-conformist.

The origins of that fear were obvious: the Indians had been overcome by means so obscene that the products, wealth and power, no matter how attained, were not to be questioned.

And with that particular factor of the mental climate went the inhibiting effects of a small community.

Edgar Lee Masters like any honest man was a tangle of contradictions: he distrusted the city and knew the value of the roof over the small town—and yet loathed the provincial mind.

Many of the people whose lives he summarises in 'Spoon River' had talent and ambition—but like fish in gaunt water they never grew to fulfilment.

Some lacked the courage, some the common sense, some the luck that is essential to all great enterprise. Others were born to fail because America aroused in them impulses that they could not express.

And for those who would be artists there was the fear that they had been born in the wrong place—no greatness could come out of an obscure town in the Mid-West.

Archibald Higbie, failed sculptor, speaks from his grave:

' ... *What could I do, all covered over*
And weighted down with western soil,
Except aspire, and pray for another
Birth in the world, with all of Spoon River
Rooted out of my soul?'

Masters himself knew better, knew the truth of Leopardi's theory that the life depicted may be crude and yet the depiction may be beautiful.

And the great strength of 'Spoon River' is that its material is generally so commonplace—and yet so fascinating.

And commonplace that life may seem—but under its prosaic texture burn passions all the fiercer for being almost smothered. People kill others or kill themselves or go mad—we see what Chekhov calls 'the silent evidence of statistics.'

Tom Merritt tells his story:

At first I suspected nothing—
She acted so calm and absent-minded
And one day I heard the back door shut
As I entered the front and saw him slink
Back of the smokehouse into the lot
And run across the field..
And I meant to kill him on sight,
But that day, walking near Fourth Bridge,
Without a stick or a stone at hand,
And all I could say was 'Don't, don't, don't.'
As he aimed and fired at my heart.

And there is the secret of Elsa Wertman, the farm girl who one day was assaulted by her boss and became pregnant. His wife is childless and adopts the baby.

But at political rallies when sitters-by thought I was crying
At the eloquence of Hamilton Greene—
That was not it.
No— I wanted to say:
'That's my son. That's my son.'

What was Edgar Lee Masters's philosophy, if one can give so solemn a word to what for him more than to most was a dragging anchor?

The answer, if there is an answer, is implied in many of the pieces in 'Spoon River' and found more explicitly in a few.

You were so human, Father Malloy,
Taking a friendly glass sometimes with us,
Siding with us who would rescue Spoon River
From the coldness and dreariness of village morality ...
You were so close to many of us,
You believed in the joy of life,
You did not seem to be ashamed of the flesh.
You faced life as it is,
And as it changes ...

And in 'Lucinda Matlock' Masters expresses his pride in the better aspects of the pioneering past. The poem is about his father's mother, a woman who bore 12 children, worked very hard, and eventually passed away as naturally as an apple falls from the tree in Autumn.

I spun, I wove, I kept the house, I nursed the sick,
I made the garden, and for holiday
Rambled over the fields where sang the larks.
At ninety-six I had lived enough, that is all,
And passed to a sweet repose.
What is this I hear of sorrow and weariness,
Anger, discontent, and drooping hopes?
Degenerate sons and daughters,
Life is too strong for you—
It takes life to love life.

But perhaps the expression of what was nearest to Masters's heart is in 'Fiddler Jones.'

The earth keeps some vibrations going
There in your heart and that is you ...
How could I till my forty acres
Not to speak of getting more
With a medley of horns, bassoons and piccolos
Stirred in my brain by crows and robins
And the creak of a windmill — only these?
And I never started to plow in my life
That someone did not stop in the road
And take me away to a dance or picnic.
I ended up with forty acres:
I ended up with a broken fiddle—
And a broken laugh, and a thousand memories,
And not a single regret.

Next to Fiddler Jones lies Cooney Potter:

I inherited forty acres from my Father
And by working my wife, my two sons and two daughters
From dawn to dusk, I acquired
A thousand acres. But not content,
Wishing to own two thousand acres,
I bustled through the years with axe and plow...

Eating hot pie and gulping coffee
During the scorching hours of harvest time
Brought me here ere I had reached my sixtieth year.

'Spoon River' is the work of one who felt that at last he had got it right, that he had found his philosopher's stone and could speak out bold and clear.

Evening Press, Tuesday, July 4th, 1978

Gilbert White — Lone Watcher

Few parts of this world have been so favoured by nature and so lovingly nurtured by man as the south-east of England. Should a Martian ever come down there, it is unlikely that he will ask to be taken up to London to the leader — he will merely beg that he be allowed stay for the rest of his days in that paradise.

Gilbert White was an exceedingly lucky man — not only was he born in that blessed corner of earth but in a parish where the essence of all its charms are concentrated.

Selborne is in Hampshire, a few miles over the border from Sussex; not very far to the north are the burgeoning towns of Basingstoke and Aldershot and Guildford—Gilbert White's heartland has changed but little since he made it famous and it made him immortal.

That is one of the great glories of this region: it changes, as all places must—but the process seems only to enhance its innate beauty.

In another circumstance White was lucky too—he was born into a clerical family in an age when that class, far more than any other, cherished learning and the spirit of enquiry.

And for most of his adult life he held a comfortable curacy in his native place—and thus had an ideal base from which to attempt the task that from an early age he seems to have seen as his destiny.

That task was alarmingly simple: he desired to discover and set down all the secrets of the non-human life in his parish.

As Newton was to the stars so he would be to the rook and the otter and the trout and the bluebell and whatever form of life, other than his parishioners, he saw around him.

81

It hardly needs saying that he did not achieve this ambition—his life's work was like that of a lone prospector who goes up a strange river and sieves the sand and finds occasional traces of gold and sometimes even a little nugget.

And all the time White was excited by his intuition that the truths he sought could not only be found but demonstrated.

He did not solve all the mysteries that tantalised him; some indeed are perplexing still—but he inspired generations by the vigour and lucidity of his intellect and the purity of his intent.

It may seem wilful of him to have confined his researches to his own parish, to a modest parcel of what is mostly pastoral land.

But this very concentration was his great strength: by taking so intimate a scale he was able to apprehend better the complexity of the web of circumstances on which life depends.

And small in area and undramatic in landscape though the parish of Selborne may have been, it contained an abundance of wild life that would astonish the uninitiated.

That is true of more than Selborne—the most striking proof of how much is hidden occurs when a lake goes dry.

No matter how familiar you are with it, you will be surprised by the abundance and variety of the life that lies revealed.

And even in a modern conurbation where all seems inimical to him, you will occasionally encounter a creature of the woods—and are as surprised as if you heard a dialect word in the reading of the news on the BBC.

And though Gilbert White's parish was small, it had marvellous variety.

In his own words the district was diversified by "a variety of hill and dale, aspects and soils ... chalks, clays, sands, sheep-walks and downs, bogs, heaths, woodlands and meadows".

On one side of this pastoral parish was a great forest; on another was a marsh where peat is dug to this day; not far to the south was the sea.

And so Selborne was a kind of cross-roads for birds—and of all Gilbert White's passions they became the greatest.

There was another reason for this: birds are more visible than most creatures of the wild—the darkness covers the activities of many of the others as the water covers those of fish.

And above all he was fascinated by those gypsies of the sky—the swallows and the martins and the swifts.

Where did they go when they left Selborne? Was it the same couple that came back to the nest they had built the previous year?

Did some of them remain behind in the Winter and, if so, how did they survive when they had no flies to eat?

The last question especially tantalised him. Now and then he heard that a swallow or a martin or a swift had been seen in the district in Winter — but invariably when he sought out the person credited with the sighting, it was to find that he had heard about it from someone else.

And he used to employ men to search in likely places for birds that might have remained behind — but in vain.

And yet despite the general belief that they all emigrated, he suspected that some few might not—if only because they were for some reason unable to do so.

And he played with the idea that they survived by immersing themselves totally in wet clay by the side of a pond or river and remaining insulated and inanimate until Spring stirred again.

That may give the impression that Gilbert White was a credulous innocent—he was nothing of the kind. The notion about hibernation in the mud rather shows his imagination and openness of outlook.

There is one especial passage in 'The Natural History of Selborne' that shows the muscularity of his mind.

He observes two swifts a considerable time after their comrades have departed. Why have they not gone too? What strong purpose keeps them behind?

He watches—and after a few days there is only one. And he begins to suspect that the remaining bird is a hen and that it is motherhood keeps her behind.

And after some time he is delighted to be proved right—one day he sees her teaching the young ones how to fly.

And eventually mother and family are missing—they too have gone south. And now he searches the church to see where they were born—and finds the bodies of an earlier clutch that had died.

And he feels he is near an important discovery — that swifts breed only once in their annual stay, unless the first clutch dies. In that simple passage you have the essence of Gilbert White.

And his observations were not only perceptive and tenacious—they were communicated with marvellous enthusiasm and in language that shows him as that rare being, a scientist who could write.

Listen to him about his beloved swifts: "It is a most alert bird, rising very early, and retiring to roost very late. And it is on the wing in the height of Summer at least sixteen hours.

"In the longest days it does not withdraw to rest till a quarter before nine, being the latest of day birds.

"Just before they retire, whole groups of them assemble high in the air, and squeak, and fly about with wonderful rapidity."

Gilbert White was one of the first to write of other creatures and of plants as inhabitants of the same world as humans—and there is in his work too a sense of all life being one.

You feel this perhaps most strongly when he writes about the language of the domestic hen.

The hen has a wide and vigorous range of language—and White chronicles most of it, accurately and delightfully.

Surprisingly he omits two elements — the complaining sounds they make when the morning outside is inclement and the sounds of contentment they utter when they have settled down for the night.

Should you ever go to let out hens for their day's roaming when the weather is cold and damp, they will let you know their doubts about your wisdom.

They give out a complaining sound that is somewhere between a groan and a whimper. Wittgenstein used to say that other creatures differ from man in having no self-pity—sometimes you doubt it.

And when you close in hens for the night after they have been out all day, if would be worth your while to wait for a few minutes and listen.

Then you will hear small sounds that are very expressive of contentment. White remarked this in swifts and speaks of "a little inward note of complacency."

And if you wait longer and listen to them while they sleep, you will be inclined to believe that they dream. If they do, they qualify to become members of Jung's collective unconsciousness.

This speculation about hens dreaming might shock orthodox scientists. But White would consider it—and even though he respected general opinion, he would query accepted beliefs when he found evidence that seemed to contradict them.

He was forever enquiring. Why, for example, is it that cats have a passion for fish that is equalled only by their aversion to water?

And how long would a pig live if allowed to run his natural course?

He had a neighbour who once kept a sow until she was seventeen — she was turned into bacon only when she was no longer fertile.

By then she had produced over three hundred pigs and her meat, you will be astonished to hear, was juicy and tender.

We are reminded of the Kerryman who when asked why he was keeping a pig far beyond the conventional time, answered that he might not get one as quiet again.

Gilbert White wrote little about his fellow human beings. Perhaps he felt that their truths were hidden under something less penetrable than even water or darkness.

He lived in an especially contradictory age. The prevailing philosophy was optimistic—but the treatment that passed as justice seemed to imply that man was a savage who needed unremitting correction.

The correction in White's time and place usually consisted in hanging the offender until he was very dead.

The great naturalist made no comment on this form of wild life. He seems to have believed that civilisation was like a wheel forever going forward.

And yet not very far from Selborne, across the sea in France where some of his parish's rooks sometimes fed, a mighty wave was building that would savagely usher in the modern world.

When the Bastille was about to fall, Gilbert White was still wondering if swallows or martins or swifts ever wintered in Britain.

He never married. Selborne was the only mistress he ever knew. It was a singularly happy union.

Evening Press, Tuesday, October 25th, 1977

Izaak Walton — happy fisherman

One of the greatest poems in modern Italian is about the eel — and if you have only the slightest acquaintance with this astonishing creature, you will easily understand why he is the hero of an epic.

It would not have surprised Izaak Walton: every chapter in 'The Complete Angler' fascinates — but none more than that about the most resourceful and mysterious of all beings on this planet.

We know something of the eel's life cycle now; in Walton's day there was only the sketchiest knowledge — and the imagination, as it always will, tended to fill in the vacuum.

In his great book be tells us of eels 'digged out of the earth with spades, where no water was near' and of how it was related that in the cruel Winter of 1125 'some eels did, by nature's instinct, get out of the water into a stack of hay in a meadow upon dry ground and there bedded themselves, but alas a frost killed them.'

The first part is certainly true, as many a ditch-maker can verify; the second sounds like an old husband's tale — but when you have been long acquainted with the eel, you will not be in haste to dismiss it.

And especially if you have ever seen their mass-going up river, you will concede that they are among the chiefest of nature's miracles.

Walton refers to it: 'I have seen in the beginning of July, in a river not far from Canterbury, some parts of it covered over with young eels, about the thickness of a straw — and they did lie on the top of that water as thick as motes in the sun.'

Those that Walton saw must have been a considerable time in the river — when they come up from the sea, moving very slowly, sometimes not moving at all, in a great wavering stocking several miles long, they are no thicker than cotton threads.

It is said that by then they have been three years travelling from their birthplace, advancing almost imperceptibly with the current and sustained only by the microscopic sac of fat with which they are born.

Had Izaak Walton known that, he would have marvelled even more: one of the great charms of 'The Complete Angler' is its sense of wonder — as Charles Lamb said: 'It breathes the very spirit of innocence and purity and simplicity of heart.'

These qualities need not be a handicap: Walton was very successful in business and also at catching fish — and would agree with Hemingway that in those fields enthusiasm is hardly enough.

And his book has more than charm: the most sophisticated angler of to-day could profit from it. Fish, and indeed humans, have not changed very much since it was first published in 1653.

One is thrilled to hear a companion of Walton's who has not yet commenced fishing for the season say: 'I have not wet a line'. It is the idiom of to-day.

And the lore then was like that now — a blend of the sensible and the mystic: Walton advises us, as any veteran angler in these islands will, to keep worms in moss and give them daily a few spoons of cream — he also believed that those of his fellows who were extraordinarily successful had some secret lure.

He tells us that camphor 'put into your bag may give the worms so tempting a smell that the fish fare the worse and you the better'. One is reminded of the belief that the smell of crushed ivy berries is irresistible to trout — later on in 'The Complete Angler' you will find that too.

One hears and reads of all these deadly lures — and can only marvel that any fish are alive at all.

But if Walton's book contained no more than a mixture of shrewd and innocent advise on angling, it would not have been reprinted scores of times — and it is significant that it lay in almost total oblivion until about the start of the nineteenth century.

By then it was infused with nostalgia: the industrial revolution was growing — and many an Englishman read of green fields and pure waters as of some wondrous land.

'The Complete Angler' fascinates the reader who has never 'wet a line' because it evokes a world where the countryside was unblemished and where life seems to have been leisurely and free.

It is in part an illusion: not everybody was as privileged as Izaak Walton — and his own life was far from being as idyllic as that of the fisherman in his book.

He lived in very troubled days; he was an outspoken royalist trading in London when King Charles was executed — and seems to have kept his head only by lying low for some time in his native Staffordshire.

Nor was his life free from domestic misfortune: all seven children of his first marriage died in infancy — as did one of the three born to him when he married again.

But he was that happy man; he had total belief in a meaningful universe — and all who knew him well have testified to his fortitude and sweetness of character.

His intervals in the countryside, then no more than a pleasant walk from his shop in Fleet Street, strengthened those traits; no doubt, the popular belief about the mental benefits bestowed by angling is well founded.

People say it requires patience — one might more truly say that it teaches it.

People say too that all fishermen are liars — they are, in fact, exceedingly truthful, but such amazing happenings occur in the pursuit of their avocation that in order to communicate them, they must tell the credible untruth rather than the incredible truth.

He who becomes 'a brother of the angle' will never lose his sense of wonder; it is not only the creatures of the water that sustain it but the birds he will see that are almost unknown to the layman.

When you are fishing bait, you are sometimes silent and almost motionless for long periods — and no more frightening to the creatures of the air than if you were a tree.

Then you may see the dipper, the bird that can swim under water, and the goldcrest, the smallest bird in these islands, so tiny that at a little distance you could mistake him for a butterfly.

You will find this sense of these side-delights in 'The Complete Angler' — you will find too an innocent zest for the simpler pleasures of life.

Walton's fisherman says to his apprentice after a long day on the river: 'I'll now lead you to an honest alehouse where we shall find a cleanly room, lavender in the windows, and twenty ballads stuck about the wall.'

We hear mention too of 'the fruitful vine, of which when I drink moderately, it clears my brain, cheers my heart, and sharpens my wit'. Indeed . . .

And then in the inn we hear: 'Give us some of your best barley-wine, the good liquor of our honest forefathers, the drink which preserved their health, and made them live so long, and to do so many good deeds.'

And every angler that ever went to the river at dawn will appreciate the following passage — as will many who have never been assailed by that desire.

'It is now past five of the clock; we will fish till nine and then go to breakfast. Go you to yonder sycamore tree and hide your bottle of drink under the hollow root — for about that time we will make a brave breakfast with a piece of powdered beef and a radish or two that I have in my fish-bag.'

The meal is 'good and honest and wholesome' but the day is long and the river is a great place to whet the appetite and they come home in the evening 'as hungry as hawks.'

And the good lady of the inn cooks for them the best of their catch — a trout twenty-two inches long, 'and the belly of it looked some part as yellow as a marigold and part of it as white as a lily'.

And after supper; they turn to the fire and drink the cup 'to wet our whistles and so sing away all sad thoughts.'

It all seems too idyllic perhaps — but such pleasures are not unknown and were especially savoured by Izaak Walton in days when to be alive and free was no small achievement.

But above and beyond all else 'The Complete Angler' is a hymn to water and the creatures that live in it.

To its author it was the primal element: all life was water in another shape. Hence came the importance of its most familiar inhabitant: 'Moses appointed fish to be the chief diet for the best commonwealth that ever was' and 'Our Blessed Saviour chose only three to bear him company at his Transfiguration — and all were fishermen.'

And it was in water that Walton saw most powerfully the life force — and most of all in the spawning urge that drives fish to overcome what seem impossible obstacles.

Salmon because they are so big exemplify this most spectacularly: 'They will force themselves through floodgates or over weirs or hedges or stops in the water even to a height beyond common belief.'

But perhaps it is the littler fish that exemplify it most compellingly, especially fingerling trout and eels big enough to have gone out on their own and yet small enough to swim in an eggcup.

If you are ever watching at some bridge where there is a floor of concrete to prevent erosion by swift water, you will witness their astonishing courage and persistence.

You will see the troutlings leap over the little waterfall where the concrete apron meets the natural bed — and repeatedly be borne back down until eventually they struggle up the gentler waters at the sides.

Up here the elvers will come too, sometimes totally out of the water but on the concrete. What makes it all the more mysterious is that the upward drive of these little fish has no immediate relation to breeding.

Izaak Walton would, no doubt venture some explanation. But he was as shrewd as he was innocent and 'The Complete Angler' contains one great corrective to those who would harp too much on angling theory: 'There was never yet a good horse of a bad colour.'

They are the words of a sensible man, which Walton surely was. He led a good life and no doubt the trout and the barley-wine played a part in his reaching 90, an astonishing age by the standards of his day.

By then he was fairly well known for his biographies, especially that of his friend, the great poet John Donne.

That book is hardly read at all now — 'The Complete Angler' becomes even more fascinating as the life depicted in it becomes more remote.

Evening Press, Tuesday, July 18th, 1978

Walt Whitman on the road

Thomas Wolfe's most celebrated novel begins: "The tragic light of evening falls upon the rusty jungle that is South Brooklyn."

And a kind of rambling piece that he wrote about a man whose hobby was to seek out new places to savour is entitled "Only The Dead Know Brooklyn".

And the great bridge that takes its name from that quarter of New York tormented Hart Crane into an epic attempt to capture it.

The very word, Brooklyn, seems to embody something peculiarly American, something akin to what Wolfe spoke about when homesick in Paris he said he longed to touch the peeling rail of the boardwalk in Atlantic City.

And perhaps if you could see into the secret history of Walt Whitman, if you could trace his soul's growth as not even the most assiduous scholar ever can, you might find something of Brooklyn there too.

He knew it when it was different — where is now a great warren was rolling farmland, and where is now a vast dock was a shore where little fishing boats went and came.

Theodore Dreiser used to say how lucky he himself had been to have grown up in Chicago when that mighty city was growing up too.

He was in at the adolescence of a great metropolis; Whitman was even luckier—he was in at the adolescence of a great continent.

You will find the yeast of those years in his "Specimen Days', the rough journal he kept off and on from early manhood.

And you will apprehend the truth of what Scott Fitzgerald wrote in "The Great Gatsby" a century later.

". . . for a transitory enchanted moment man must have held his breath in the presence of this continent, compelled into an aesthetic contemplation he neither understood nor desired, face to face for the last time in history with something commensurate to his whole capacity for wonder."

That is a typical Scott Fitzgerald statement — its music lulls you into accepting it without question.

Its main thesis is true—but not everyone who contemplated the great physical and spiritual outback could be said to do so without understanding or desire.

Walt Whitman, if the statement had been made before his time, would have seen himself as an exception, as someone who was fated to be the voice of that sense of wonder.

He had, as many great men have, an element of the charlatan in him—but his concept of himself as poet-prophet of the New World was not without substance.

When he conceived that idea and why we will never know — perhaps in the lives of many men there are watersheds that remain hidden even from themselves.

Whitman's early life was hardly that of one who was imbued with a sense of destiny—he may have perceived his star but it hardly held him on a steady course.

He worked at many jobs, not, one senses, like some later writers who did so during Summer vacations in their young manhood and made it all seem very glamorous much later in potted biographies on the back cover of the Penguin edition.

Whitman seems to have done so from necessity—and even though his ramblings from place to place and job to job may have constituted an excellent apprenticeship for the role of Great American Poet, it was hardly a voluntary one.

He seems to have been like a ship caught in a storm with a defective rudder and a dragging anchor — and the storm was the immensity and unuttered poetry of America.

Thomas Wolfe knew that experience too three generations later. He spoke of seeking a key and a door; he expressed the inchoate feeling of many that America had been betrayed. He expressed too the hope that is the birthright of a new country—"We are lost, but we will be found."

Whitman's America was more like a great page on which only a few diffident marks had yet been made.

At lease he saw it that way—it is one of the peculiarities of American thinking that so many of its great men have been able to come to terms with the fact that their nation had been founded on genocide.

Whitman was well aware that the Indian had not very long before hunted and fished in the Brooklyn of his youth—somehow he managed to skip over that uncomfortable truth.

So did Thomas Wolfe: what the two have in common is an unshakeable sense of America's greatness and a vast hunger to write an epic proportionate to it.

Wolfe hardly succeeded: in his frenzy he tended to equate sheer outpouring with expression. And about his work, great though it is in parts, there is a terrible sense of things felt but imperfectly said.

Whitman, on whom most would more readily bestow the name of "mystic" is the solider artist—and it is possibly because of his very different kind of early life.

You might say that Wolfe was always a wordsmith: writing was more or less the only job he ever knew. But "wordsmith" is a deceptive term because it implies a testing in an objective world of your work.

Whitman had been farmhand and compositor and printer and had a more intimate knowledge of how life is underpinned. .

'Mystic' is a much-abused word. One is reminded of Yeats's idea that love, far from being blind, perceives the high beauty that is hidden from other eyes.

'Rhetoric' is an abused word too, so much so that it has come to mean raising the pick high without sinking it deep.

Wordsworth declared war on it—but his best poetry is not altogether written in the humble language he professed to deem best.

And when you are faced with the infinite wonder of an unfolding continent, you are likely either to be stricken with silence or to sing out loud and clear.

'Leaves of Grass' sings. Was there ever a more marvellous title? It implies that the poems were not the result of long and painful labour but a kind of natural growth.

And that seminal book came at a time in Whitman's life when there was little evidence of greatness in what he had published.

He had written indifferent prose and sentimental verse. The wanderings of his life seemed without direction. It seemed that he would reach his fortieth year without having made the faintest mark.

And suddenly here was a voice as powerful as Niagara, as embracing as the Mississippi. Emerson said that when he read 'Leaves of Grass', it made him rub his eyes.

It was as if for more that three decades Whitman's spirit had been scattered in weak streams all over a great plain. Now the plain became narrower and gave way to a slope—and the streams came together in a powerful river.

The preface to 'Leaves of Grass' is a kind of charter for American poetry—it is not without bombast, but neither is it without great truths, if not always too happily expressed.

In that manifesto, as innocent as Wordsworth's, Whitman is like someone who has fallen overwhelmingly in love and must tell the whole world all about it.

And like the lover, he continually invokes the name of the beloved—in this case it is the mighty land into which he was born. "The Americans of all nations at any time upon the earth have probably the fullest poetical nature."

And the essence of that poetry was to him in the heartland of his youth. In all his writings he comes back again and again to it: "Brooklyn of ample hills was mine".

And like so many of America's great writers his love of his own few square miles of boyhood arena is wedded to a great hunger to know the world of which it is a part.

Thomas Wolfe was a country boy, or at least a provincial—it is not altogether untrue to say that he killed himself trying to know and express the greatness of New York.

Walt Whitman felt that hunger too but not so overwhelmingly—New York was a much simpler place in his day and he knew it fairly well.

The vastness that tormented him was more that of the great continent unfolding inland—his 'Song of the Open Road' is a mighty declaration of intent.

> *"Afoot and light-hearted I came to the open road,*
> *Healthy, free, the world before me.*
> *The long brown path before me leading wherever I chose."*

And among his earliest writings you will find many little jottings about the joy of movement, though then it was sprung from his simple journeyings on the Brooklyn ferry.

And though it is a hundred years since Whitman's odyssey, "Leaves of Grass" is still whispering. Its echoes go on — because it is so powerful and because what it said is so true that it is perceived by kindred souls.

"So in America when the sun goes down, I sit on the old broken-down river pier watching the long, long skies over New Jersey and sense all that raw land that rolls in one unbelievable huge bulge over to the west coast, all the people, dreaming in the immensity of it . . ."

92

There is rhetoric in that passage too—but it is legitimate rhetoric. It echoes Walt Whitman and Scott Fitzgerald — it is part of the final passage in Jack Kerouac's 'On The Road'.

What you hear in it is something peculiarly American, an especially powerful example of what Herder taught to the young Goethe—that great poetry is always the result of a people's spirit.

Evening Press, **Tuesday, August 15th, 1978**

Knocknagow — No Rural Idyll

Whitman's great ambition was to become America's 'national' poet, the man in whose work the people of the mighty nation would find their life resoundingly expressed.

He hardly succeeded. The man whose verse left the most traces in the popular mind was Longfellow.

It is curious that so gentle and conventional a writer should have been embraced by so energetic and adventurous a people — but there are parallels elsewhere.

If there is such an entity as the Irish mind, it would not be too harsh to say that a certain distrust in the goodness of the universe is part of it—yet the writers that we have most taken to our hearts were both kinsmen of Candide.

Oliver Goldsmith is one. Do not ever doubt that—it is not something you can prove by statistical analysis, but if you are in the habit of keeping your ear to the ground—in other words, indulging in the odd quiet drink — you will know.

In this island there are people walking abroad who will quote him at the slightest provocation and when songs are being exchanged in a pub, beware the innocent bystander in the corner—he is waiting his chance to give a little passage of about 300 lines or so from 'The Deserted Village.'

Charles Kickham is the other great favourite: for every one Irish person acquainted with Leopold Bloom there are perhaps ten familiar with Mat The Thrasher.

You may think that Kickham is known mainly in rural Ireland — but 'Knocknagow' came out in a new edition recently and although the majority of Irish people now live in towns and cities, it was at the top of the list in last week's best sellers.

93

Books sometimes become best-sellers for the wrong reasons: they are dishonest concoctions designed to prey on the immaturity that lurks in all of us.

'Knocknagow' is as honest a book as ever was distilled from a mind's fermentation—and in the hundred years since it came from the presses has been very widely read. In vulgar parlance, it must have something.

What is that something? The facile answer is that it provides an idealised picture of Ireland—and even though we know in our hearts it never was, we find comfort in the dream.

That is in part true—and when you talk to people about the weather of your youth, you tend to remember only the finer days; so too, our picture of 'Knocknagow' is a montage of happy scenes.

There is Barney Wattletoes being asked by the young gentleman from England 'Have I much time to dress?' and answering: 'Lots of time, sir. Only if you don't hurry, you'll be late.'

And Billy Heffernan, the king of the local flute-players, finds himself before experts from the big world outside and though daunted, performs nobly and earns their warm acclaim—and a neighbouring girl cannot restrain herself from jumping into the middle of the floor and shouting: 'Three cheers for Knocknagow and the sky over it.'

And there is the most famous scene of all — when Mat The Thrasher takes on a local aristocrat at throwing the sledge and compels him to say: 'Your equal is not in all Europe.'

And it is a fair bet that if you questioned a score of people who had read 'Knocknagow' in their youth and not opened it since and asked them was it a happy book, at least fifteen would answer: 'It is.'

At best they would be only half-right. 'Knocknagow' is curiously akin to one of those days so familiar in Ireland—that begin so full of sunshine and that decline gradually into a mixture of darkness and light.

There is no scarcity of sentimental passages in it—but the general picture that Charles Kickham paints is not idealised.

Nor do all those people live happy ever after. Captain French, Mat's opponent in the sledge-throwing duel, returns from the Crimean War with only one arm: Jemmy Hogan, one of the greatest of the local hurlers, returns with only one leg.

And since the day of the local festival when Jemmy hurled, there has been great economic change. The Captain asks Mat: 'Do you have hurling still?' 'The hurlers are gone' replies Mat — and he looks around 'upon green pasture fields with scarcely a house within view.'

It is not pretentious to say that 'Knocknagow' sometimes seems less a novel than a sprawling thesis that has for themes the caste pattern of Irish society and the evils of a system under which most of those who worked the land did not own it. Both themes are intimated early.

In the very first pages we meet Henry Lowe, a young English gentleman who is spending Christmas as the guest of his uncle's principal tenant.

When he awakes on the first day of the season of good will, one of his earliest thoughts is that he is a land agent in embryo—and that his room is on the ground floor.

And this train of thought gave the holes in the window-shutters a new interest to his eyes. It was Christmas—but it could still be open season for agents.

Soon we meet the hero of the book—that is if it has a hero. Mat Donovan is tall and brawny; he excels in all kinds of work as a farm labourer and can turn a hand almost to anything.

'He is a magnificent specimen of the Irish peasant,' says Mr. Lowe to Mary Kearney, his host's daughter. And she says: 'Let us wait until you hear him speak. You will be sure to hear something out of the common from Mat The Thrasher as we call him.'

It is plain that Mary admires Mat—it is even plainer that the chances of the two coming together like Gabriel Oak and Bathsheba Everdene in 'Far From The Madding Crowd' are extremely remote.

And what indicates the rigidity of the caste system is that Mat himself would not dare to entertain such a thought—he is the kind of man who would die for his country but would not question the social order.

Mat in a way is a tragic hero: he is splendidly gifted but he lives in a society where these gifts are unlikely to be over-abundantly rewarded.

His nickname is indicative—it was not bestowed on him for his athletic prowess but because he never met his match at wielding a flail. 'As a consequence he was in great request among farmers from October to March.'

Indeed, he is in demand all the year round—for, among other things, 'his superiority as a ploughman was never questioned.'

And it is one of the ironies of 'Knocknagow' that the best man in the locality to work the land is a man who has hardly any land at all.

The first time we meet Mat away from his admirers he is melancholy and discontented. The melancholy is induced by love-sadness. The discontent is related to his economic status. And the two sources are not unconnected.

The girl that will not let his mind be at ease is a little above him in the social ladder — and he says to himself: 'Sure I know a poor man like me has no right to think of her.'

And later when we see into the mind of his beloved, we find out that she admires him greatly but almost rules him out as a suitor because he is only 'a poor labouring man.'

Thus in what may seem to be an innocent manner, Kickham lays bare some of the reality that lies obscured by the happy picture of the lads and lasses dancing on the village green.

And, even if unconsciously, the most revealing passage in this huge book is the description of the Donovan household at their evening meal.

Mat lives with his widowed mother and his sister Nelly. All three are well above the average of their class at managing their affairs—yet the supper is hardly a feast.

It consists of an amplitude of boiled potatoes, a few small leeks, and a single salted herring. 'All three helped themselves with their fingers to the herring, which they took in minute pinches, as if they were merely trying how it tasted.'

This is hardly a picture of a bold peasantry living off the fruits of the good earth—it suggests rather that they were existing on the thin of the land.

Nor is Billy Heffernan, the great flute-player, pictured as a kind of Con the Shaughraun, the life and soul of every fair and dance and wedding.

Perhaps it is true of any big book to say that its most memorable passages are those that came from deepest in the writer's heart—if so. Charles Kickham was no stranger to loneliness.

Billy works very hard. He is a bogman in the fullest sense. Everything in his house seems to have come out of the peat — even the table and the chair are made from bog-oak.

But hard work does not make him happy. As he sets out for Clonmel with his creel of turf, he meditates: 'The road is lonesome and the house is lonesome and the bog is lonesome. And the main street of Clonmel is the lonesomest of all.'

And as he comes home that evening, he is smitten by the fear that he will find something wrong—even though he has not 'father or mother, sister or brother, wife or child to awaken that feeling of dread.'

Kickham is trying to express something more than personal loneliness, something that is part of the land itself.

One is reminded of the passage in 'The Graves at Kilmorna' where Myles Cogan, the Fenian leader, is returning from his long imprisonment.

He is thrilled to be back, even if apprehensive — but suddenly on the train between Cork and Mallow the loneliness of the Irish landscape 'smote him.'

Charles Kickham and Canon Sheehan lived in times when the effects of the Great Famine were most palpable.

There is the loneliness of places where human life never was; there is the greater loneliness of places where it had once teemed and now has ebbed. Kickham and Sheehan knew that loneliness at its worst.

'Knocknagow,' far from being a rosy picture of mid-nineteenth century rural Ireland, is a gently-outlined portrait of a society and a culture in decline.

In the last chapter even the hurlers are no more. Tillage is no longer the great source of money—grass is now king and another wave of emigration has been set in motion.

And Kickham makes the moral very clear—in the end of the book we see that humble Billy Heffernan is as much its hero as Mat The Thrasher.

Mat is doing well—and his beloved has married him after all. He is now a kind of farm manager—but Billy is more independent.

Billy is the tenant of a stretch of moorland that few wanted—but he has it at a fixed rent and the lease is for as long as grass grows and water runs.

Many of his neighbours who had looked down on him have been evicted to make way for sheep—but Billy lives on in his little republic.

By the book's end he is married—and getting twenty barrels of oats to the acre out of ground that not long before had been wild bog.

In its own quiet way 'Knocknagow' is a complement to James Connolly's 'Labour In Irish History.'

Evening Press, Tuesday, August 29th, 1978

Hopkins — Maker of Language

It has been said—and by eminent critics—that Gerard Manley Hopkins did violence to the English language; it is a judgement as sensible as that which condemned the revolutionary painters of the nineteenth century for disrupting the 'laws' of harmony and perspective.

Nor was it a concidence that he was their contemporary—it is rather an illustration of the oft-ignored truth that all art is essentially one.

It is not pretentious to assert that there are affinities between Hopkins and Van Gogh.

> *"Summer ends now,*
> *Barbarous in beauty, the stooks rise*
> *Around; up above, what wind-walks! What lovely behaviour*
> *Of silk-sack clouds! has wilder, wilful-wavier*
> *Meal-drift moulded ever and melted across skies?"*

Surely that is akin to Van Gogh's 'Road With Cypresses' where he paints the sky and the moon and the trees and the hedge and even the home-going workmen in a way that they seem to dance to his soul's music.

And Hopkins is like Cézanne too in his endless speculation about the nature of his craft, a theorising both took to such excesses that you wonder how either wasn't inhibited from ever creating anything at all.

And Hopkins is like both Van Gogh and Cézanne in that for most of his life he worked without companions, without material reward, and without the only fame that matters—the understanding of his work by kindred spirits.

His tragedy was to some extent different from that of the two painters: their work was rejected by the public—his was not published in his lifetime on account of the rules of the order in which he was a priest.

In that he was probably lucky—it is unlikely that he would have been understood. And rejection might have turned him away from poetry altogether.

It may seem strange that these three geniuses (as they were, if the word has any meaning) should have been so neglected—but it isn't strange at all.

And the reason is bound up with the nature of the judgements that for so long condemned all three.

People in the sense of a consenting mass do not make the 'laws' of language or painting—the truth is that there are no laws, only accepted modes.

And words and word-orders are all the creation of individuals, the more verbally sensitive whom the rest of the tribe follow. So, too, it is with painting.

And a time comes in the development of the language of poetry and of painting when all seems to have been explored—for the new adventurer there is no outback, only gleanings.

He must strike out boldly and perhaps a little blindly into fresh territory in the potential of language.

Hopkins did not do violence to English — such a charge assumes that it has a natural order and texture that cannot profitably be broken.

He attempted to do totally what William Blake and John Clare had done in fragments—to speak a language so fresh that his poems would be like walls in which every stone would glow with individuality.

And what drove him on that quest? The impulse was no different from that behind the cave drawings of early man—the desire to make an image of things and thus lessen your fear of them.

It is even more than fear—it is the instinct to possess and understand that which torments you, to know it in your own terms.

You can see that even in animals—a dog that has just killed a rabbit will sometimes roll over and over on it and make noises of the utmost satisfaction.

It is unlikely that Hopkins often allowed himself such noises—he was too acutely aware of the uniqueness of things, of Blake's 'infinity in a grain of sand' and Van Gogh's 'poetry in a few furrows in a field.'

Things and events impinged on him fiercely—his poetry was his attempt to express them in a way that would give him some ease. That was all—but it was an enormous all.

It would not suffice for Hopkins to create the verbal equivalent of the drawings of the animals on the cave walls—though in a sense his work has an affinity with them.

He would retain their freshness and energy but would use exceptionally sophisticated techniques to capture as much as possible of the totality.

That is one of the most remarkable aspects of Hopkins: his poetry may seem as spontaneous as a fountain springing up out of the earth—but the water has been brought and shaped by a cunning system of pipes concealed under the ground.

And this seems to present a paradox: one cannot doubt the wisdom of Leopardi's dictum about all great things being born out of heat—or of Ezra Pound's 'There are no rules for writing poetry—only the music of the line.'

But mental impulses, no matter how powerful, do not readily translate themselves into a poem or a painting or a piece of sculpture or music.

The mountain is sighted—but before you can look down from its summit, you may have to undergo an arduous slog during which you sometimes doubt the wisdom of the journey.

That some of Hopkins's manuscripts resemble battle-fields strewn with the corpses of words should surprise no one—the final versions read naturally, the product of patient and sweaty toil.

And far from doing violence to the language he restored some of its parts that had long been neglected, at least in conventional poetry.

There is another kind of poetry, that which belongs to the generality of the people rather than the dominant social group.

But much of it tends to fall over the cliff into oblivion—the only substantial body that has survived in English is the balladry of the Scots border country.

And there you will find a freshness and an immediacy that affect you in a way akin to Hopkins's poetry.

Hopkins was not the first to be aware that the language of English poetry was in danger of becoming too Latinised — Cowper and Wordsworth had known that too.

But his experiments went farther; he restored not only many half-neglected Anglo-Saxon words — he also laboured to get poetry away from the conventional rhythms and nearer to those of speech.

It would be idle to pretend that he was always successful—but when he is, you feel that language is entering a new dimension.

"I wake and feel the fell of dark, not day;
What hours O what black hours
This night . . ."

That little fragment from one of his sonnets illustrates another aspect of his work—the boldness of language is found more in expressing the external world than in articulating concepts.

You can see that very clearly in the famous sonnet 'Thou art indeed just, Lord' — in which he expresses despair at what he sees as his own spiritual impotence.

In it the language of the concepts is, as always in Hopkins, strongly individual — but when he comes to capture the life of the fields, it takes off.

It is a strange poem: it is harrowing in its sense of frustration—and yet one cannot help feeling that it was written not without pleasure.

"See banks and brakes
Now, leaved how thick! laced they are again
With fretty chervil, look, and fresh wind shakes
Them . . ."

He was a great word painter—it is a reflection of his seeing things as fiercely individuated.

And if things are so individuated, then surely the most complex of all creations on this earth, man himself, must in every person be the most individuated of all.

And hence comes the need for a language that will at once capture the storm in the speaker's mind in a way that will satisfy himself and be intelligible to the listener.

In this ambition Hopkins did not always succeed: when you find 'O Hero savest' for 'O Hero that savest', you are not in the territory of genius but of self-indulgence.

But such instances are rare. Those who complain of how difficult it is to understand him are often looking for something that is not there at all.

He is a highly sophisticated poet—but only in the legitimate cunning he employed to express a childlike vision.

Again one thinks of Van Gogh: people failed to understand him too—because what he is saying is so simple.

Perhaps the best key to Hopkins's vision is that little anthology piece 'Pied Beauty.'

It begins:
"Glory be to God for dappled things—
For skies of couple-colour as a brinded cow;
For rose-moles in all stipple upon trout that swim . . ."

He goes on to talk of
"Landscape plotted and pieced—fold, fallow, and plough;
And all trades, their gear and tackle and trim."

He is talking about individuation—but he is talking too about the relation between things and words, man's most valuable and exciting creation.

And this brings us back again to the origins of language. Poets in their various ways make a language—it becomes the language of the people in the sense that they accept or reject.

The earliest poems were made up of only a simple word—the man who first called the salmon 'salar' captured its essence.

But it would be misguided to think of Hopkins as one who so delighted in verbal felicity that he was inclined to forget the world out of which words spring.

The verbal felicity is wedded to high seriousness—nowhere is that clearer than in 'Felix Randal,' perhaps the most satisfying of all his poems.

Its subject was a farrier who had been one of his flock when he ministered in Liverpool. There the poverty and squalor shocked and depressed him— again a parallel with Van Gogh.

And Hopkins like Van Gogh could see the dignity and even the nobility in the midst of the misery.

He tended the dying farrier and gave him Extreme Unction 'our sweet reprieve and ransom.'

The poem ends:

"How far from then forethought of, all thy more boisterous years,
When thou at the random grim forge, powerful amidst peers,
Didst fettle for the great grey drayhorse his bright and battering sandal."

There, as so often elsewhere in Hopkins, and in Van Gogh and Cézanne, you feel that something has been expressed as never before. That is genius.

————————

Intimate Life in a Countryman's Diary

A man keeps a diary. Why? I mean a real diary—one that tells the truth about his life as honestly as he can distil it.

But did any one ever keep such a diary? Is not the urge to tell it 'as it happened' far more likely to produce a novel?

Well, perhaps Samuel Pepys did. You like him because he was not afraid to be seen now and then in an unflattering light.

But he was fairly certain that nobody would read his daybooks in his own lifetime. He did not leave them lying around—and the code would have baffled the casual finder.

But that complex code meant too that he could not easily read back over his own words.

Why then did he write? The answer is probably simple enough: he had the urge to record—and in his time and circumstances a diary was the most convenient form.

And that urge to record is bound up with time's passing—you do not want it to go by uncontested. You hunger to capture some little part of every day.

Several generations after Pepys a similar urge possessed an Irish schoolmaster—the diary he left behind is not as celebrated but is no less fascinating.

Pepys's work has great intrinsic merit—but gains mightily also from his position close to the heart of government and from his love affair with London.

He was at the centre of things in a town that was at the centre of the world.

Humphrey O'Sullivan spent most of his days in a village in the south-east of Ireland and could hardly be said to have ever been close to the mainstream of Irish life.

He made occasional trips to Dublin—by coach it was only about seven hours away—and sometimes stayed for several days. But the city has little place in his diary.

Dublin was then suffering from the loss of its parliament — but there seems to have been a deeper reason for O'Sullivan's failure to record even its most superficial aspects.

He seems to have been one of those people whose faculties work fruitfully only in their native habitat.

O'Sullivan like Gilbert White was so affected by his own heartland that it almost became an element outside which he shrank.

He could spend a week in Dublin and record it in a few lines—an hour's stroll by his local river might yield several pages.

And yet his Callan is as absorbing as Pepys's London. In a way it is even more so. No country has been better chronicled than England—and the impersonal parts of Pepys's Diary are a kind of bonus.

O'Sullivan gives us a window onto a world about which our knowledge is tantalisingly fragmentary.

The formal histories of his Ireland are mainly taken up with the campaigns for Catholic Emancipation and Repeal—of the intimate life we hear little.

O'Sullivan was deeply involved in politics—but it is as a private man that we mainly come to know him in his diary.

But the private man was continually making notes on the general life — and they are an invaluable supplement to the broad sweep of academic history.

For instance, we tend to think of The Famine as a sudden calamity—it wasn't.

It was more the dreadful culmination of a pattern that had for a long time been unfolding.

In all that generation before the terrible years of the forties the ingredients had been gathering.

Early marriages and subdivision of land and reliance on one crop were a fatal mixture—the result was nearly as inevitable as ever human tragedy could be.

As early as the April of 1827 O'Sullivan was recording: "The famine is all over the countryside — Callan's own paupers — fifteen hundred persons—are reduced to misery.

A collection is being made for them by Lord Cliften, the priest and the minister of the parish, the chief magistrate and the Callan merchants. But it won't last . . . for long, God help them."

A few days later he wrote: "I hope to God the people's hearts will soften. Three hundred families in Callan are starving."

It is not clear who 'The People' were in this context—but elsewhere in the diary we find that the small farmers were the best friends the poor had.

The local middle class — the merchants and the big farmers and the professional people — did not seem to worry unduly because many of their neighbours were dying of hunger.

And certainly O'Sullivan himself, even though a compassionate man, did not allow the misery of the poor to inhibit his enjoyment of life.

His diary is full of exceedingly zestful passages about eating and drinking, mainly at the house of the parish priest.

Our image of the Irish Catholic priest of the early nineteenth century pictures him as a poor hunted saint-hero saying Mass in caves and on mountain sides and with a price on his head.

Father Seamus Hennebry, parish priest of Callan, was not quite like that. In fact, to put it simply, he did himself well.

"Four of us had a meal with Fr. H. We had a boiled leg of lamb,carrots and turnips, roast goose with green peas and stuffing, a dish of tripe boiled in fresh milk, port and punch and tea."

So wrote Humphrey O'Sullivan in the September of 1829. It is not easy to see what Father Seamus had to gain from Catholic Emancipation.

And that little meal was only a mere snack compared to others lovingly described. And the guests sometimes numbered twenty — with the fair sex well represented.

There would be singing and dancing — and a variety of drink that makes one suspect that Ireland in some respects has progressed backwards.

And certainly if Yeats's 'The good are always the merry' is true, then Fr. Seamus was not for from being a saint.

But the begrudgers are always with us — and a local priest abetted by a Callan merchant brought certain charges against Fr. Hennebry.

O'Sullivan does not specify the charges — but they are not too difficult to guess. The Bishop heard the case — and the diarist was one of those who gave evidence.

Father Seamus walked out without a blemish on his name and the feasting went on better than ever.

Alas! The good priest would seem to have done himself rather too well — and paid the penalty.

One morning in the January of 1834 he fell down as he dressed himself and when the doctor arrived, he was very dead.

Apoplexy was the enemy—nobody seems to have been surprised. And we are told that young and old wept for him.

And the good priest proved as generous in death as he had been in life — he left his considerable store of money to the poor of Callan.

And if he seems a contradiction — a gourmet and gourmand in the midst of famine — it is because we tend to forget that life is infinitely complex.

Even in the darkest times it is impossible to prevent gaiety from breaking out.

Certainly in Humphrey O'Sullivan's diary the poor and the near-poor of South Kilkenny are not always on their knees.

Their love of dancing speckles his book — and those who deplore the manners of to-day may be surprised to find that the young girls of 150 years ago did not shun the pubs.

Here is part of O'Sullivan's entry for Easter Monday, 1827: "A bright sunny calm morning, at mid-day the young girls and young men taking their eggs and drinking in the hostelries. Evening — and the hostelries still full of young people."

Sometime someone will attempt to trace the history of neo-puritanism in Ireland.

O'Sullivan had not the slightest guilt-feeling about drink. One of his little asides on the subject is revealing.

Father Padraig Grace, an Augustinian, dies at ninety. He had lived "a life of sanctity. He was an urbane pleasant virtuous man. Nobody ever saw him drinking to excess."

And it might be interesting to attempt to trace a connection between the temperance movement and the growth of tuberculosis.

But it would be wrong to deplore the temperance movement—the pity was that eventually it contradicted its own name and took a good idea to excess.

That it was needed is only too plain in O'Sullivan's diary. He regularly complains about the crazy drinking of the poor and the loutish behaviour it induced.

And in his heart, even though he worked to improve their lot, he despised them: !'It's no harm to call them 'mob' for they are the froth of the lake-dwellers, bog-dwellers, and dirty mountain-dwellers with no respect or manners."

That outburst was caused by their keeping him awake until three in the morning after the day of the eggs and the drink — and can be excused. But its wording comes from good old Gaelic snobbery.

Humphrey O'Sullivan was not lacking in common sense and got out of school teaching as soon as he could — he entered the drapery business and prospered.

He became an employer — but has left a record of only one of those who worked for him. She was his servant, Margaret St. John.

Her surname may be a corruption — but it is more likely that she was from that strange class, the poor of English descent.

Elsewhere in the diary you will find spailpins with very English names. Anyhow, Margaret or Peggy as her master later called her occupied a strange position in the O'Sullivan household.

It is clear enough that her position was rather like that of an American black in the old South.

She was paid very badly even by the miserable standards of the day — and had to work in the fields as well as in the house.

She came shortly before his wife died, left when she gave birth to a child, but soon returned.

The last entry concerning her is only slightly cryptic. "Spent the night very foolishly in the company of Tomas Toibin and with M. St. J."

That was in the summer of '34. For the March of the following year there is an entry: "Ellen Tracey came as a maid to me today at four shillings a quarter."

There are other signposts too that indicate that Humphrey was all too human—but they are of only trivial interest compared to his general observations.

And over and above everything else is the question of the land, the real key to the history of nineteenth century Ireland.

The absentee landlords were bad—but their native agents were worse. And in O'Sullivan's time a new class was growing up — the small man who had acquired money and was buying land.

In Kerry a generation later the first man to be murdered in the agrarian war was a local peasant who had bettered himself and bought a small farm that included a few tenants.

The tenants feared that he might evict them — a hired gun was their remedy.

In O'Sullivan's South Kilkenny there was a similar killing — but with more provocation: the new landlord had turned five families onto the road.

O'Sullivan while out walking one day, met one of the evicted — "a poor tall thin ragged barefooted woman, . . . weeping bitterly".

Seemingly she is living in the open near her old house. Her lamentation is concrete and harrowing.

"It was I set those potatoes but it is he who will dig them; . . . my spinning wheel is in the ditch, my table on the fence, my pot in the undergrowth . . . "

There is a true voice of the dispossessed peasant, especially in the Irish context.

Not surprisingly, the new landlord did not live to enjoy the potatoes. Two men literally swung for him — three more were deported to Australia.

And there they found plenty of comrades from the neighbouring island. Nineteenth century Britain was no paradise either.

On the sixth of January, 1831, O'Sullivan records: "In 16 counties of England hay and corn stacks are being burned by the workers, who are being hanged and tortured, and deported to Botany Bay.

"It is dire poverty that is forcing the people of Ireland and England to rise up against the law of the country".

That dire poverty can be glimpsed from later entries in the diary.

You meet children out gleaning — and people stealing the loose potatoes off the ridges — and burning thistles and dry cowdung for fuel.

And yet at a time of terrible poverty O'Sullivan estimated that there were 100,000 people at Kilkenny races.

Perhaps the most valuable element of the diary is how O'Sullivan almost unconsciously shows you the seeds of the Great Famine, the most important turning-point in modern Irish history.

The causes were many — but the biggest was very simple and is still with us: a country may have an abundance of food — but the dominant classes do not always agree to it being distributed free.

Evening Press, Tuesday, March 23rd, 1980

Lowry — Will I endure?

At present in the Royal Academy there is on exhibition a mighty collection of paintings that are loosely dubbed post-impressionist — earth, as William Wordsworth said in a different context, has not anything to show more fair.

The front of the great building is decorated with a line of streamers that advertise the treasures within — and in the middle are three that carry prints of famous self-portraits.

They are of Van Gogh and Cézanne and Gauguin — and the selection is not without irony.

The atmosphere in the Royal Academy speaks of success and the good life and the triumph of beauty and truth — but Van Gogh and Cézanne and Gauguin were terrible failures both in their own view and that of the public.

Van Gogh despite popular belief did not commit suicide — no more than he cut off one of his ears: he wounded himself with a revolver in a classic cry for help — but he was in poor health and this wound would prove fatal.

It isn't true to say that he died unknown: he was greatly respected by some of his fellow-artists — but he had sold hardly anything and depended for his sustenance on his brother's charity.

Cézanne never had to worry much about money: he came from a wealthy family and had a private income — but his life too was fretted with misery.

What harrowed him most of all was that hardly anyone understood him: he knew that he had made great discoveries — but the world wouldn't listen.

He was like an explorer who had seen wonderful things but would not be believed.

And what wounded him most of all was the publication of 'L'Oeuvre', the novel by Emile Zola that portrayed him as a failure.

Zola and he had been boyhood intimates in Provence and had gone to Paris together to conquer the big world. Zola had succeeded — in a sense. Cézanne had failed — in a sense.

Cézanne's last years were enormously unhappy. He had returned to his native place and had hardly a friend there — certainly there was nobody to whom he could talk deeply.

He was not greatly respected by his neighbours: they thought it unseemly for a man of his affluence to be unkempt and roughly-dressed and they did not discourage their children from throwing stones at him.

But he worked away doggedly, painting at all hours and in all weathers until death brought him ease. His life was about as romantic as that of a ragpicker.

Gauguin's was little better: he suffered as if the fates had singled him out — and his last years were mottled with disillusion and illness.

Why did three such giants fail almost totally to win appreciation in their lifetime?

The obvious answer is that they were originals: when we say a man comes before his time, we usually mean that he wasn't lucky enough to find a contemporary critic who understood him.

Acceptance of an original is usually a gradual process: at first he is seen as at best misguided and at worst insane — the mass of men are desperately conservative.

And when someone stands up and expresses startling truths, the reaction is like that of a herd of cattle that turn on a newcomer.

The established critics are usually the worst: they behave as if the tenets on which they have founded their life's work are being sabotaged.

But then someone among them with unusual perception and honesty and moral courage speaks up and acts as a kind of interpreter for the newcomer.

And gradually the old skin is peeled off people's eyes—and they begin to understand.

We can all see now the greatness of Van Gogh and Cézanne and Gauguin — and we flatter ourselves that had we lived in their day, we would not have been as obtuse as their powerful contemporaries.

But we forget that we are heirs to three generations of re-education—that is part of the meaning of L.P. Hartley's dictum about the past being a foreign country.

Van Gogh and Cézanne and Gauguin knew well that they could not hope for general acceptance — but they failed beyond their worst fears.

If they had a consolation, it was in knowing — or at least occasionally sensing — that they were great and would endure: they must have had such moments — otherwise they could hardly have carried on.

When you look at their work in the Royal Academy, in an atmosphere that speaks of easy living, you can hardly help dwelling on the contrast.

It is a splendid gallery — but you feel that it is in the wrong part of London.

Around it are streets dedicated to the glossy and the luxurious—you walk through them quickly lest you be charged with trespassing.

And the few bars nearby are mostly in hotels whence you fear you would be evicted like the man who came without the wedding garment.

But some of those whose pictures are in the exhibition at the Royal Academy are at home in such an environment.

They were extremely fashionable in their day — they were trendy and glossy and had the knack of pushing themselves forward.

One of their contemporaries went to the other extreme. He disliked being called an artist—or perhaps pretended to do so: he was 'a man who painted' and his studio was a 'workshop.'

He was fond of saying: "You cannot keep a bad man up." His own career indicates that the converse may be true too.

Most of his fashionable contemporaries scorned him—but his exhibition at the Royal Academy two years ago was thronged for the ten weeks of its duration.

The crowds queued in the courtyard outside—and they came from all walks and runs and stumbles of life: L.S. Lowry had achieved something very rare—popular and critical success.

By then he was two years dead: and, in a way, that was typical of his life: most of the things he wanted came to him late.

He was a man with an infinite capacity for self-mockery— and would be enormously amused by the cult that has grown around him since his death.

His image now is that of the wise fool whose private life was a mystery but who though a simple uneducated Lancashire man, saw his industrial environment with the eye of a mystic.

The truth is that there was very little mystery about his life — but because he was a puckish fantasist, he led the curious up all kinds of garden paths.

He was seen as uneducated because he had left school at fifteen and was given to saying that he had never passed an examination.

The latter feat was less meritorious when you considered that he had never sat for one — and leaving school early meant little in an age of stern self-improvement.

He also liked to give the impression that he was a kind of hermit whose life had been a long struggle against poverty and loneliness and critical neglect — but it was not so at all.

Lowry came from a comfortable family, had adoring parents, and never knew the humiliation that is the worst aspect of poverty.

He was born in 1887 in Old Trafford, a semi-suburb of Manchester that has since become famous for reasons other than his birth.

In later life he pretended that he was hopeless at school — but great men tend to exaggerate when indicating the depths from which they have climbed.

For the first four years of his working life he was a very efficient clerk in an office of chartered accountants — from there he went to an estate agency and stayed for over forty years.

He started at twenty-two shillings a week — it wasn't bad money in 1910. And the job gave him remarkable freedom.

He worked for an amazingly enlightened firm—and was free from the tyranny of rigid office hours: if his work was done, that was that.

His work was to collect rents in mainly working-class parts of Manchester—for most of the week he was out and about. His duties back in the office were not onerous.

Whether by chance or by design he was in a job that could not have suited him better: it gave him freedom—and great insight into the lives of ordinary people.

He came to love the working class in a way that had nothing to do with political dogma or sentimentality—he admired their toughness and their honesty and their humour.

He found that they had no self-pity: they paid the rent cheerfully when they had it—when they hadn't, they invited him in for a cup of tea.

And though he hardly realised it at the time, they were to be the mine in which he found his purest gold.

Lowry had been drawing since early childhood and had been attending night classes in art since he started working — but with no great ambition or no clear sense of direction.

In later life he would tell many different versions of his Damascus, of how one day some simple incident suddenly made it clear to him that his mission was to express the life of the industrial world he knew around him.

It was all great stuff for those who like life to be dramatically scaffolded — but the truth was at once more complex and more prosaic.

Lowry was not a pioneer of art based on the life of the British industrial working class: it was all around him in his more formative years.

There is no scarcity of theories 'explaining' Lowry's preoccupation with working class life. Maurice Collis in "The Discovery of L.S. Lowry" quotes the words the artist spoke to him: "I was with a man and he said 'Look' and there I saw it. It changed my life."

And if you believe that, you would believe anything. It would be wrong to call Lowry a liar: the world is divided into two classes — those who are slaves to surface truth and those who have imagination.

A more sensible explanation is that he was influenced by contemporary writers, especially Stanley Houghton, whose 'Hindle Wakes' opens to a backdrop of mill chimneys.

Oddly enough, his official biographer, Shelly Rohde, does not mention D.H. Lawrence or 'Sons And Lovers,' a great novel wrought out of a mining village by a man who was almost his exact contemporary and who grew up not very far away.

But influence is often overrated: it is usually no more than a shower that stimulates what is already there. There were other — and probably more compelling — factors that turned Lowry down his own especial road.

One was his obsession with the problem of freedom. How free is man, even in such a comparatively democratic country as Britain? Lowry saw most of his fellows almost enslaved by ill-rewarded toil. Economically he was reasonably free — but for many years had to care for his mother.

And he was fascinated by those people who for some reason or another had fallen off life's wheel and were condemned to scrabble on its outskirts, the people you see hoarding the heat in public libraries and big railway stations.

When he met a beggerman, he would wonder what had happened to him, what had been the breaking point. And those he came to know he found far more interesting that his own neighbours.

And Lowry saw the mill worker as both trapped and to some degree broken. He felt life had played a foul trick on them—but he loved them for their endurance.

He envied them a little too—for though he never lacked for friends and received a fair amount of critical acclamation quite early in his career, he was in a sense a very lonely man.

Underneath all the secrecy and the myth-making with which he fought off those who were curious about his private life, one very important truth could not be hidden: he was so shy with women that a great part of existence passed him by.

In his younger days this was construed as coldness—it was, of course, the very opposite: he loved the forest so much that he feared it.

When he was old (but in fact he was never old—only in the calendar) fame and wealth came to him. And there was no lack of adoring young women, a few of whom became his disciples.

But Lowry, like Manet, was wont to say that it had all come too late. By then he was obsessed with one especial fear. Would his work endure? Was the stuff of greatness in it? Few doubt the answer now.

I, Edward Thomas

They say that Robert Frost converted me into a poet: the story goes that we first met one night at a literary party and that he took me aside and gave me advice that changed my whole life.

He had read some of my pieces on rural themes—and felt that the stuff of poetry was in them.

It seemed to him that I was searching for something that prose could never bring.

And he advised me to attempt to put my thoughts and observations and intuitions into the shape of verse — and that it might act on them as water on parched plants.

And the story goes on to tell how I followed his counsel — and at last my desert began to bloom.

Well, it's a nice story — but life rarely has such cleanly dramatic turning points.

Robert Frost was a good friend to me — but I was attempting verse before ever I came to know him.

The encouragement he gave me came from example rather than advice. He had created great poetry out of a seemingly undramatic background — and I felt that in poetry I might express the forces I felt in myself.

For years I had been aware of those forces and guiltily aware.

Ever since the end of my days at the university I had been writing for a living — and postponing what I thought of as my 'real life'.

My life, God knows, was real enough: I had a wife and family — and commissioned books and literary essays kept me as busy as a fox in spring, but far less secure.

And yet I felt that it was not real at all — I was like a man doing hard labour in prison and saying to himself that he would make up for all the lost days when he would be released.

That is one of the insidious things about being a writer: your attitude towards time—and indeed towards people—tends to be cavalier.

You secretly believe that the only true reality is what you create — and that no matter how unpromising your life, you will win through in the end.

And so I toiled on, sometimes deceiving myself that I was a writer — but more often smitten with remorse because I was not.

I was tormented by feelings of depths in myself that I might never know.

And in my worst moments I feared that my real self had been smothered under the accreted self — and would never again come to the surface.

And yet in the innermost tabernacle of my mind I knew that my real self still lived — occasionally I could hear it, like water flowing underground.

And I had written some prose pieces that I knew were good and would endure — in them my mind's excitement had through a mixture of luck and patience found a happy mould.

And yet that too tormented me: because I had done that little, I felt I should have done far more.

I felt like a man who had come to an inn one day and met a lovely girl behind the bar and talked with her and sensed that she liked him — but was afraid to go that way again.

And I used to assuage my guilt by saying to myself that the drudgery of providing for my family left me with little time or spirit.

But that was not wholly true — I was like a man who has been so long in prison that he dreads the day of his coming out.

And my dreams of being a poet had too vulnerable a fabric — I could never read an essay on poetry without feeling my own inadequacy.

What tune could I play to match the organ of Milton — or Tennyson's harp? At best I could finger out a few tunes on the tin whistle in a sheltery corner of a wood.

And they wouldn't even be jigs or reels or hornpipes — only slow airs fumblingly uttered.

And yet I always knew that some day I would go back to the inn.

That lovely girl had left me too disturbed — I would rather fail than not know if my passion was returned.

And for me that lovely girl was a mystic compound of skies and fields and trees and streams and moors and downs and roads and paths and hedges and birds and fish and animals and people — it was the old-long-lived-in-heart-of-England.

It haunted me. And I felt that if only I could capture some little fragment of it, I might gain a little ease.

And in my bolder moods I imagined myself being to that world in words as Constable was in painting.

It was a strange ambition in one who was of Welsh blood and who had grown up in a great metropolis.

These, perhaps, were the very reasons why the south east countryside of England overwhelmed me.

When you are a kind of outsider you are more piercingly aware — and you long to prove yourself a worthy member.

And so I explored that world patiently and yet hungrily — like a rook in a long dry spring.

I came to know its skin better than most people ever did — and all that grew and dwelt on it and the things that man had made and the skies above it.

And I knew it was my heartland and that whatever good work I would ever do would be rooted in it.

In me was a deep and lasting excitement — but I was afraid I might not be able to translate it.

And yet the simplest effort brought release — even though I knew that sometimes I was no more a poet than is the lover when he utters the name of his beloved.

But there were other times when I felt a little satisfaction, felt that my words had some kinship with the reality.

The first poem I ever completed was — it seems to me now — wildly ambitious.

It was about a lonely inn that I had come across in my travels — lonely but yet heavy with a sense of oldness. I tried to express its uniqueness.

While I drank
I might have mused of coaches and highwaymen,
Charcoal burners and life that loves the wild . . .

And sometimes a simple scene would so impress itself on me that I knew it would come out as a poem—if I had the patience.

That was one thing I learned—that poetry doesn't flow from you.

Creating a poem is like building a drystone wall — it is a task where you choose and reject and choose until you know that the words are right.

People have said to me that Adlestrop is a magic poem, that it must have been born like a leveret, born fully-fledged.

And yet the manuscript is like a battlefield — you keep on changing until a sense of rightness declares itself.

Yes, I remember Adlestrop
The name, because one afternoon
Of heat the express-train drew up there
Unwontedly, it was late June
The steam hissed.
Someone cleared his throat
No one left and no one came
On the bare platform, what I saw
Was Adlestrop — only the name.
And willows, willow-herb and grass
And meadowsweet and hay-cocks dry,
No whit less still cloudlets in the sky.
And for that minute a blackbird sang
Close by, and round him, mistier,

Farther and farther, all the birds,
Of Oxfordshire and Gloucestershire.

That little mouse of a poem is no more than the impression of a moment — but it is a bold little mouse and gave me heart.

For years and years I had been keeping notebooks, filling pages with what I saw and thought on my travels.

Now I began to put some things into verse — I felt like a painter with a sketchbook.

And for some strange reason there are moments in your life that you know you will remember forever.

There are those rare moments when you see the present and the local invested with the fertility that generally comes only with the perspective of time.

And then you feel at one with the world — and you sketch confidently.

A man might go on forever thus, recording the lucky moments — but if you are a poet, there is no resting place.

Sometimes you envy the painter — because in a sense he works outside the world of concepts and human relationships.

And human relationships are the most tortuous of all to express. You find it almost impossible to get the stones to make a sound wall.

And yet in a poem you can say something to someone dear to you that you could not otherwise say.

I wrote poems to my children and to my wife — and felt that it was the only way I could honestly talk to them.

Especially to my wife I tried to explain myself.

And you, Helen, what should I give you?
Many fair days free from care
And heart to enjoy both foul and fair,
And myself, too, if I could find
Where it lay hidden and it proved kind.

And always I returned to the fields and the woodland paths — in that world I felt most myself.

I had grown up in a London suburb, a shapeless place where life was ragged and rootless.

And I longed for a world that had a roof over it, a world that had grown like a tree.

I came to know the life of the countryside — and longed to feel myself a part of it.

And do not imagine that I saw it sentimentally — I did not.

I knew harsh springs when I heard the rooks praying for rain. I pitied them — and yet envied them their sense of community.

And I knew how hard life could be for even the shrewdest and most energetic creatures of the woods and fields.

A fox might forage all day for his vixen and cubs — and come home at nightfall with only a half-decayed crow he had plucked from a stake in a cornfield.

And yet he knew the kind of security I would never know — he was free from self-doubt.

And I envied the workers on the land too, even though I knew that their lives were for long stretches hard and monotonous.

But they knew their sweet moments — these were like the little flowers that grow only in moorland.

And they were part of the fields. They had helped to shape them and the fields had helped to shape them.

And what had I been for all my working life? I had been a spinner and weaver of words — no more.

And then the war came. I enlisted. I found comradeship. And I began to find myself — more and more.

And because I felt myself more a man, I had more faith in words.

And I wrote a poem that I know will endure. That was in the Spring of 1916, when I was on leave from France.

It tells about my conversation with a ploughman. It begins:

As the team's head brass flashed out on the turn,
The lovers disappeared into the wood . . .

I sat on a fallen tree. The ploughman said it would have been removed if his mate had not gone to war.

The second day
In France they killed him. It was back in March
The very night of the blizzard too . . .
Then the lovers came out of the wood again.
The horses started and for the last time

116

I watched the clods crumble and topple over
After the ploughshare and the stumbling team.

And I was reminded of something I had once written — and with which Robert Frost would certainly have agreed:
"When a poet writes, I believe he is often only putting into words what some old man had puzzled out among the sheep in a long lifetime."

Evening Press, Tuesday, August 14th, 1979

Chronicling the birth of a Great Nation

Sherwood Anderson is like a pure and vigorous river—and if you have once known him, you will always return and by his banks renew your sense of wonder.

And as you marvel at the sweep and freshness of his waters, you may remember what Canon Sheehan said: "It is the throbbing of Niagara that gives America its mighty energy."

Sheehan's statement is shorthand for saying that the soul of a people is profoundly affected by the physical nature of the land they inhabit.

And America generates an enormous sense of power — and from it emanates not only a feeling of the unknown but of the unknowable: it has a permanent outback.

Sartre wrote an eloquent essay about its fascination for the French, those people who are as deeply rooted as a sycamore tree.

Peter Maurin, the great socialist philosopher, for instance, came from a family whose ownership of a little vineyard goes back to the eleventh century.

And Sartre himself is a kind of urban peasant: his heartland is geographically small and its boundaries are clearly defined.

But the American has yet to shed his frontier mentality—he is still moving westward in his covered wagon, hungry to see what lies at the other side of the mountain.

He still yearns for earthly Edens. Part of his dilemma is that he has come to a promised land — but it has implanted impulses in him that cannot readily find fulfilment.

And his soul has been distilled out of movement. He is a child of the greatest river of immigration in history.

And movement is the yeast in many of the books he has accepted as his classics, as the most faithful mirrors of American hope and experience.

117

In the gallery of his mind he sees a boy on a raft drifting down the Mississippi and a demented sea-captain lusting after a giant whale and a people whose land had blown away trekking towards the green pastures of California.

And yet two of America's most seminal books seem like landlocked pools that a great overflooding river left behind as it returned to its course.

But 'Walden' is deceptive: it is a product of movement too—Henry Thoreau is standing back for a little while to see if the wagon-caravan is on a true course.

And when he speaks of his pond, he is indulging quietly in the art of understatement—it was a considerable lake. That he should call it a pond indicates America's vast scale.

Thoreau did not hide away from the main stream of American life: he retired for a little while so that he might measure its flow.

In 'Walden' you are away from the turnpike and the market-place, but you are aware of a raw and powerful nation in the making.

In 'Winesburg, Ohio' you can feel its pulse: Sherwood Anderson chose as his measuring point a little town in the heart of the Mid-west.

'Mid-west' in strictly geographical terms is a misnomer—but nobody quarrels. It seems to summarise perfectly an aspect of America's spiritual development.

Scott Fitzgerald saw the Mid-west as a kind of living museum of the pioneers' America — in 'The Great Gatsby' he contrasts its innocence and simplicity with the corruption and sophistication of the east.

His judgements may have been subjective — but Anderson would seem to agree: his most famous book is a kind of elegiac survey of a lost America, of the time when every small town 'had a roof over it'.

'Winesburg, Ohio' is a basket of short stories; some of them are connected with others—what they all have in common is their tegument.

And even though it seems a casually-written, loosely-arranged book, it imparts a powerful feeling of unity.

And though some of the people in it you can never forget, its profoundest impression is of a time and a place.

The place is a town that is still a little island in the fields; the time is about the end of the century—when the forces that shaped modern America were beginning to surface.

Winesburg is small enough for all its inhabitants to know one another—and yet you can live and die alone there. And some people do.

They are the ones who will not or cannot make treaties with a world that does not answer to their dreams.

It marches on while they listen to a different music. They become outsiders — not only lonely but distrusted. The frontier mentality might be fading but it demanded conformity—it still saw the Indians lurking on the edge of the clearing.

And the street corner and barber's shop senates might seem no more than confluences of ignorance — but you felt that at a lynching their leaders would be the first men to put a hand on the rope.

America was as 'lovely a land as ever lay out of doors'—but the dominant element in reshaping it was a coalition of brutishness and greed.

The effect of that vast and beautiful country on the sensitive was to arouse dreams almost impossible to realise.

Its effect on some others was to create fear and distrust — and a wish to subjugate it, to possess it in the destructive sense.

'Winesburg, Ohio' in the main tells a familiar story—it chronicles the struggle between ideals and the forces that are hostile to them.

But because the arena is the America of a century ago, the battle takes on a rare intensity.

The first story in the book is about a 'fat little old man' who had come to Winesburg twenty years before but who did not think of himself as in any way a part of the life of the town.

The people of the place know nothing of his antecedents. They do not know even his real name. They have christened him 'Wings' because of his extraordinarily expressive hands.

His only friend in Winesburg is a young reporter named George Willard.

We meet the old man on an evening when loneliness has smitten him to the marrow.

It is Autumn and from his door he can see a wagon filled with berry-pickers returning from the fields. "The berry pickers, youths and maidens, shouted boisterously."

Only George knows his story. For years he had been amidst young people—he had been a gifted and passionate teacher.

But almost inevitably his passion—and his endlessly expressive hands — had been misinterpreted. He was brutally beaten and driven from the town in Pennsylvania where he taught.

And in fact, when we meet him, he is not an old man at all—he is only forty. But he is like a tree that has aged because it is in a hostile environment.

He is a broken man—a poet who had tried to write in a language not yet grown sensitive enough for his needs.

When we leave him, the rumble of the train that took away the day's harvest of berries had passed. The night is again silent. He remains an exile—even in that little mid-western town.

119

The most celebrated story in the book is about an exile too — even though on the surface he is in the very middle of life.

Ray Pearson is a farm hand. He has a wife and a flock of small children and his shoulders are rounded "from too much and too hard labour".

When we meet him, he is working at the corn and "in a sad, distracted mood".

"If you knew the Winesburg country in the fall and how the low hills are all splashed with yellows and reds, you would understand his feeling."

His workmate, a young man half his age, is in 'a woman scrape' and asks him whether he should marry.

Ray does not know what to say: society has conditioned him to one answer — his instinct to another. He himself had been in a similar 'scrape'—and had married.

He feels that life has tricked him—that what was fine and bold in him has been betrayed. And he feels that whatever he tells his workmate will be a lie.

When we leave Ray, it is coming on towards night and "even the little clusters of bushes in the corners by the fences were alive with beauty".

He has no time for contemplation. 'His little sharp-featured wife' is ordering him to go to the town for groceries — "You're always puttering. Now I want you to hustle."

And as he goes on his errand, he thinks of things that hadn't come into his mind for years.

"He hadn't wanted to be a farm hand . . . he would go to sea and be a sailor or get a job on a ranch and ride a horse into Western towns, shouting and laughing . . ."

That story, 'The Untold Lie', illustrates one of Anderson's central themes.

He believed that a great many people are not their true selves — that they have lost their way in life and become diminished by circumstances and by accreted ideas and attitudes.

Occasionally, as happened to Ray Pearson on that Autumn evening, something occurs that momentarily lifts the skin off their eyes—in most cases it soon resumes its dulling power.

Sherwood Anderson's own life was a dramatic illustration of the true self escaping from the conditioned self.

He was in his forties before he started to write—and the transition, in his own telling, was in the manner of Paul Gauguin.

He was a successful paint manufacturer who felt that for a long time he had been walking on a dry river-bed and wanted to wet his feet again.

And so one day on a sudden impulse he abandoned office and factory and family and set out for Chicago with only a few dollars in his pocket.

The truth was that he had worked so hard at his business that he suffered a shattering mental breakdown—or so it seemed.

When he recovered sufficiently to take honest stock of himself, he knew that the real cause was that for years he had been fleeing from his true self.

He had realised the American dream—now he set out to realise his own.

And because Anderson had been a very shrewd businessman, it would be strange if his greatest book contained only portraits of bruised people.

Successful businessmen are in part prophets—they foresee developments and cater for them.

And one of the glories of 'Winesburg, Ohio' is the tremendous sense it gives you of a growing nation.

Superficially the book is static—but the great maker of Anderson's America, the railway, is threaded through it: you are never very far from the whistle of a train.

And you see Winesburg, the little shopping town, taking its first steps into the modern world.

And you will find too the beginning of America's obsession with the machine.

And many of the inventors were men of the land: the huge fields were so laborious to work by hand that the search for mechanical aids was intense. For those early Americans the machine meant freedom.

Anderson chronicles all this (less in 'Winesburg, Ohio' than in some of his novels) — he had a most acute awareness of workaday life.

So, for instance, had Sinclair Lewis who sometimes seems to be compiling catalogues for Sears Roebuck — but Anderson was aware of a great deal more.

He had a marvellous sense of place—and when you walk with him at night down a little path at the back of the immigrant labourers' houses, you feel that you are in at the birth of America.

And when he describes the feeling of being on a just-deserted fairground, he is measuring the unmeasurable — but you know that he is right.

"On all sides are ghosts, not of the dead but of the living people . . . The place has been filled to overflowing with life. It has itched and squirmed with life. And now it is night and the life has all gone away. The silence is almost terrifying."

When you meet such passages, you understand Malcolm Cowley's comment on Anderson: "There are moments in American life to which he gave not only the first but the final expression."

'Winesburg, Ohio' is by far and away his best work. One hesitates to call it a book—it is more like an apple-tree.

And as he was moved by unfulfilment, so too he was fascinated by the sense of budding.

The last piece in 'Winesburg, Ohio' tells about young George Willard going away out into the big world and hoping to be a writer.

We leave him as the train goes west for Chicago. His home town is now 'but a background on which to paint the dreams of his manhood'.

And thus the first and last pieces in the book end with the train moving away. Sherwood Anderson was a cunning poet.

And you will always go back to him, as you will to Cézanne. Both were fumblers—but because they sought something very big.

Evening Press, Tuesday, September 24th, 1985

In a house of Art

Do you remember Wordsworth and the daffodils? Of course you do.

He was wandering lonely as a cloud when all at once he saw a host of daffodils.

And he went out of his mind in the best possible way.

The point of all this preamble is that recently I enjoyed a similar experience.

It was the morning after The Republic of Ireland's clash with Switzerland.

Dis I say 'clash'? They met with all the force of two pillows colliding.

And on the morrow morn I had a tiny glimmer of how Napoleon felt after Waterloo.

And so after a none-too-hearty breakfast I went for a walk, by the river of course.

Patrick Kavanagh was right about water: it is the best psychiatrist in the world.

And so I ambled along by The Aare, a wide stream so pure that it reminded me of Ireland long ago.

And as it was my last day in Berne, I behaved for once as a tourist unashamed.

In simple language, I set out to visit its more famous sights.

The Cathedral and the House of Parliament came first - because they were nearest.

The Cathedral is 15th century gothic. It has marvellous stained glass windows depicting The Last Judgement.

The House of Parliament is famous for its terrace — whence you can see The Bernese Alps.

All this was worthwhile. Indeed I began to feel slightly virtuous.

I felt, however, that something was missing — and soon the answer surfaced.

I hadn't gone to the art gallery. Was there an art gallery? It was an unworthy fear.

122

Some people — perhaps many people — tend to think of the Swiss in terms of cowbells and milk chocolate and numbered bank accounts.

You would be as close to reality if your images of Ireland consisted mainly of shamrocks and shillelaghs and leprechauns.

Anyhow I decided to ask a policeman — the catch is that you should search long and hard to find a policeman in Berne.

And so I asked a man who looked as if he knew about such things. He did.

He assured me that it was just across the river. I followed his instructions — it hadn't moved.

I wasn't expecting a great deal: so much of the world's great paintings are in France, Britain, The Low Countries, Italy and the United States.

The Art Museum in Berne is remarkable in its atmosphere: unlike so many public galleries it is more a house than an institution.

The rooms are small; you feel at ease there.

I expected to see the work of local artists, worthy people whose fame never crossed the frontier.

The first room bore out that expectation: it is devoted to the work of Ferdinand Hodler, a native of Berne Canton.

He lived from 1853 to 1918. And as far as I could find out, he did most of his work in Switzerland.

He is a painter of enormous power, a man not afraid to ask big questions of himself.

His work is melodramatic in its tendency to overstatement but there is no doubting his integrity.

His most ambitious pictures are symbolic but the one I remember most clearly is a realistic painting of a wood cutter.

You see him with his axe at the end of the backswing: the sense of power and energy is overwhelming.

Another local painter has a room to himself too. Albert Anker lived from 1831 to 1910 and there can be no doubting is love for Switzerland.

He is the homeliest of painters — he celebrates the simple life.

One of his bigger pictures shows a rural festival: some people would deem it sentimental; it isn't — it expresses the intuition that people can occasionally be happy.

His still lifes are the quintessence of domesticity.

The bread and the ale are eminently eatable and drinkable.

And the interiors hint at a kind of comfort that seems especially Swiss.

I was glad to meet these strangers and deemed the experience to have made my day.

Little, as the man said, did I know: the best was yet to be.

I wandered into another little room and suddenly came face to face with Paul Cézanne.

He is among my special heroes but I haven't seen many of his paintings in the flesh.

There is only one in the Berne Museum but it is among his most powerful.

It is a self portrait in which he is wearing a kind of flat cap.

It is an almost savage portrayal of loneliness, the loneliness of one who cannot find his place in the world.

Cézanne was unsentimental to a fault: he had no illusions about the romance of art.

And by the time he painted that portrait he had few illusions left about life.

His boyhood friend, Emile Zola, had publicly judged him a failure — in the novel 'L'Oeuvre'.

Time has reversed the contemporary judgements on both — but Cézanne saw 'the wedding feast of the world' pass him by.

So did Vincent Van Gogh. There are two pieces of his in the museum.

One is a rather cluttered still life. The other is powerful: it is a first draught for one of the faces in 'The Potato Eaters'.

There are pieces by Picasso there too — but old Pablo never spoke to me.

I have always suspected him of being flash. It is perhaps a ludicrous suspicion — but that's how it is.

My old friend Gustave Courbet is there too. It is hardly surprising — he spent some time as a political refugee in Switzerland.

He is represented by a typically bold still life depicting two fish.

The caption tells us that they are trout — I suspect that they are salmon.

And Edgar Degas is there. He has two pieces, both about horses.

One depicts racehorses. The other shows two work-horses, lying down tired.

There was a time when I thought that his depiction of racehorses was stylised.

When he shows them in the distance, they seem unrealistically 'poetic'.

But one day at The Curragh I discovered that he was right.

The field for the Sweeps Derby were away over across the plain, waiting to be loaded: it could have been a Degas painting.

And one time too I used to think that L. S. Lowry's little people were his own creation.

But when you see a moving crowd from a certain height and a certain angle, you begin to understand.

Auguste Renoir is well represented in the Berne gallery.

His pictures there are of he kind we most associate with him — great voluptuous women, romantically expressed.

Alfred Sisley is there too with a typically poetic landscape.

And so is Camille Pissarro, that sturdy opponent of trendy nonsense.

And there too is another old friend, Claude Monet, that great painter of water.

The Berne Museum is a lovely little gallery. It has a nice café — between one thing and another you could spend the day there.

And just below flows The Aare, the river with the faint green caste that comes from melting snow.

I was happy there — and for the umpteenth time an old question surfaced.

Why can visual images be so powerful? And some, such as old photographs; may have no claim to art.

Old photographs affect us probably through a kind of nostalgia.

The power of painting comes from a different source — but it is tortuously difficult to define.

T. S. Eliot could be infuriatingly pretentious but had good moments too.

And his fingerpost towards a definition of art seems to indicate the right way: he spoke about taking the here and now and making it rich and strange.

It is easier for the painter than the writer — and probably easier for the composer too.

Some painters — Van Gogh is an example — didn't seek for Chekhov's pearl in a dungheap.

And even late in his career he was still obsessed with people. Cézanne seemed hardly aware of people at all.

He was in thrall to the mountain near his birthplace. He drew it and painted it again and again.

What was he attempting? I suppose the obvious answer is the right one: he was trying to capture it.

You will hear various explanations of the cave paintings — but when our ancestors depicted certain animals, I suspect that they, were striving to 'capture' them too. To 'capture' in this sense means to make an image of something in the hope of diminishing its power over you.

That, of course, is only one aspect of art: Albert Anker painted his Swiss scenes because they delighted him.

The winter of Lawrence's life

Perhaps the most revealing of all D. H. Lawrence's writings is an article published in The New Adelphi Magazine in the summer of 1930.

He wrote it the previous winter. He was then 43. Within a year he was dead.

It is a strange essay, sad yet vibrant. And unlike most of his magazine pieces it is almost totally free of the ranting infallibility that could make him so obnoxious.

It is not, of course, totally free: Lawrence, the prophet and giver of general judgements, was never far away.

It is called 'Nottingham And The Mining Country'. When he wrote it, he was living in the south of France and so ill that he must have known the end was near.

This essay is like a little homecoming: Lawrence, the world wanderer, returns in spirit to his heartland.

And it is so calm and almost humble that it seems to mark a new stage in his writing.

And it reads as if written for himself: there is none of the striving for effect that tempts even the most honest writers.

It also casts a strange light on 'Sons and Lovers'.

The novel that first brought him to public notice is palpably autobiographical; its heroine is his mother—the father is not flatteringly portrayed.

We see the mother as sensitive and ambitious for the children; the father is coarse, dishonest, and something of a drunkard.

Sixteen years later when he wrote the essay about Nottingham and the mining country, he saw things differently.

In a remarkable spiritual turnabout he more or less rejects his mother's values and accepts his father's.

And we find that Lawrence, whose marvellous understanding of women is the keystone of his great novel, 'The Rainbow', seems now a kind of superior male chauvinist.

How his father reacted to 'Sons and Lovers' isn't recorded: he couldn't but have seen it as other than an act of low treason.

By the time it was published. Lawrence had left home. And he never lived in Nottingham again.

The essay in The New Adelphi is like an apology, a kind of confession that is also a form of penance.

And it is also an act of homage to the miners and thus to his father.

Lawrence was too delicate to go 'down pit' but this didn't prevent him from acquiring a tremendous understanding of the miner's life.

And in this essay there is nostalgia for his lost world and half-spoken regret that he had never been fully part of it.

We are witnessing a familiar dilemma: the best observers are often those who, to some extent, are outsiders—and they tend to become sentimental about that world of which they cannot be part.

The essay on Nottingham is hardly sentimental but it tends to be romantic.

It says little of the hardship. There is no mention of nystagmus or silicosis nor are we told that most miners of that era were physically old by forty.

George Orwell in his famous essay, 'The Price of Coal', went to the other extreme.

He painted so grim a picture that the romance was lost. That essay is sentimental—if in a strange way.

Lawrence's picture may be flawed but it is nearer the total truth.

He expresses powerfully the physical intimacy of men who work together at difficult and dangerous jobs.

This comradeship breeds a kind of love that has nothing to do with homosexuality, a kind of love that is almost mystical because never articulated.

Lawrence saw his father and his fellow-miners as men who lived by instinct, men in whom the rational element lay dormant.

He overstated the case—by this stage of his life he wished this to be so.

He had flirted with many philosophies, made a few fatuous experiments at establishing ideal communities — and in general was tired of what he called "isms".

And he envied the men who had lived without being sicklied over by the pale caste of thought.

And, of course, there was much to admire in the miners: they were brave and hardy far beyond the average; they were proud of their hazardous calling.

And because they saw so little of the world above the ground, they were intensely aware of it.

And they knew the fields better than those who worked in them.

That is true of more than miners: I could say the same about the men who lived in the lanes in my own town a generation ago.

They knew the fields with an intimacy that few countrymen could match.

This was especially true of the ex-soldiers, men who had learned to forage in strange parts of the world.

The miners of Lawrence's youth were good foragers too. They knew where to find the best wild fruit and mushrooms and how to capture a rabbit.

And Lawrence believed that their appreciation of beauty was much finer than women's.

He believed that, for example, the men loved flowers for themselves; the women saw them as property, as an extension.

It is an extraordinary concept, as untenable as the Welsh half-joke about the miners and the piano in the front room: the woman wanted it as a decoration; the man wanted it so that his children could learn to play.

Lawrence's statement about the flowers is extended into a general condemnation of woman's role in the miner's life.

He sees them as the eternal naggers, always complaining about money and about jobs that needed to be done around the house.

No doubt they complained: the miners were poorly paid—and the welfare state was still only a Fabian dream.

Lawrence was, of course, well aware of the penury but his argument goes beyond it.

He sees the women as the draggers down, the enemies of all that is fine in life.

His thesis evokes 'The Last of the Summer Wine' but it isn't funny.

And you wonder why he has changed so much: Miriam, the girl who encourages the aspiring artist in 'Sons and Lovers', is a long way from Nora Batty.

And yet for all its straying from common sense, this is a profound essay.

It tells you about the miner's life in a way that a sociologist could hardly express.

"My father loved the pit. He was hurt badly more than once but he would never stay away.

"He loved the contact, the intimacy, as men in war loved the intense male comradeship of the dark days.

"They didn't know what they had lost till they lost it. And I think it is the same with the young colliers of today."

That last sentence refers to The General Strike, a struggle that Lawrence doesn't specifically mention.

The strike, despite its name, was essentially a miners' strike. That was among the reasons why it failed.

Lawrence saw the young colliers as different from his father's generation.

He believed that they had lost their ability to live by instinct: they had been lured by material prosperity—and beaten down.

It is a curious argument and Lawrence elaborates it in a way that is surprising in a man who was 'a great passer of examinations'.

"In my father's generation, with the old wild England behind them, and the lack of education, the man was not beaten down.

"But in my generation, the boys I went to school with, colliers now, have all been beaten down."

Among the factors that beat them down he lists "board schools, cinemas, clergymen," and "the whole national and human consciousness hammering on the fact of material prosperity above all things".

This is romanticism bursting its banks: even today, over half a century later, the miners are still badly paid.

The impulse behind this strange essay is Lawrence's longing for the world of his childhood.

In his case the nostalgia was especially acute: when he left Nottingham, he lost something which he never recovered.

His best work is rooted there — 'Sons and Lovers', 'The Rainbow', and a few great short stories.

The later novels all contain memorable passages but they lack the unity of impulse that in those two books enables him to transcend his failings as an artist.

The later novels are cobbled — and badly cobbled.

'Women in Love' is an exception. It comes near greatness but it is spoiled by passages so silly that they achieve comic status.

F. R. Leavis, Lawrence's great champion, described him as "the greatest creative genius of our age" — but in a sense he wasn't creative at all.

He couldn't write well when he was inventing: he was like a painter who could depict only what he saw.

Evening Press, Tuesday, June 3rd, 1986

A strange venue for a strange encounter

1988, if the lunatics who are messing with nuclear power allow it to arrive, will be a centenary year: it will mark the beginning of the world boxing championship.

The first generally recognised world heavyweight title fight was that between John L. Sullivan and Charlie Mitchell at Chantilly in France.

It seemed a strange venue but at the time the authorities in Britain were under pressure from the do-gooders to make pugilism illegal in fact as well as in theory.

And so the American champion and the British champion squared up in a ring built on Baron Rothschild's gallops.

The bout was fixed for the eleventh of March; it wasn't a clever date for an open-air fight — not surprisingly if was rained off for two days.

I said the ring was 'built': of course it wasn't — the practice then was to make a rectangle with stakes and ropes; the earth was the floor.

The occasion saw what was probably the smallest attendance ever at a title fight; not counting the seconds and the referee, forty-one souls watched. No doubt the Baron didn't want every Tom and Dick — not to mention Harry on his estate.

In other respects it was a strange encounter too: Sullivan weighed two and a half stone more than Mitchell, who was really only a middle-weight — by our standards.

And they fought in driving sleet and rain: why the action wasn't transferred indoors to a barn is hard to understand.

It was a classic instance of fighter versus boxer. Rounds then were not measured by time; when a fighter went down, the round ended. He was given a minute to recover.

A round could last less than a minute or could go on for a long time. This first generally-recognised world title fight ended in the 39th round.

By then both men were exhausted and hardly able to strike a blow — but Mitchell was the fitter and the more likely to survive.

And so some of Sullivan's backers entered the ring and asked the referee to declare a draw. He did — and then departed the scene with commendable haste.

And so the stylish middleweight from Birmingham went to his grave claiming that he had been robbed.

The fight had lasted three hours and ten minutes — and the sleet and rain hadn't helped.

Perhaps you imagine that the two gallant men were carried off to hot baths and hotter drinks and then to a party to celebrate the epic.

It didn't happen quite like that: the pugilists had just put on their clothes when the gendarmes arrived; most of the spectators scarpered — but Sullivan and Mitchell were sitting ducks.

And their traumatic day ended with a night in the cells. Next morning they were in court. Sullivan put up a defence and was fined £200. Mitchell conceded and got a token fine.

And so the first world championship bout was well and truly put into the records.

The two met again but in different circumstances; the next episode concerns a man whose name resounds through the history of pugilism. Jake Kilrain sounds like a character out of Ernest Hemingway — and certainly he had a great name for a fighter.

There was a steely hardness about it — and it wasn't surpassed until the coming of Tony Zale.

Kilrain was as hard as his name: he figured in a bout that was often recalled as the fiercest of the bare-knuckle era. That too tool place in France, in an island on the Seine. Jem Smith was his opponent; they battled for two hours and thirty minutes.

They might have gone on for as long more but it got so dark that they could hardly see each other.

Now a man called Richard K. Fox enters the story. He was the owner of a weekly paper named The New York Police Gazette.

It was an impressive title but slightly misleading: The Gazette was a lurid publication which confined itself to sport and the theatre.

Fox specialised in dreaming up sensational headlines and then concocting stories to follow them. The art isn't unknown to-day. And he did well — so much so that he got delusions.

He came to hate Sullivan because John L. hadn't gone out of his way to ingratiate himself — and his great ambition, apart from making money, was to dethrone him.

And so after Kilrain's great fight in France, the bould Fox became his sponsor.

They challenged Sullivan. He declined. The Gazette came out screaming that John L. had forfeited — and that Kilrain was now champion.

It all seems like a gimmick but it wasn't: There was no purpose in gimmicks in those days because gate money hardly mattered.

Anyhow, a fight was arranged between the champion and the alleged champion. It was fixed for the fourth of July 1889.

Sullivan was now turned thirty. He was living better than most: exhibitions were bringing him in money just as they do for golf and snooker stars now.

The difference is that to play golf or snooker you need to be reasonably sober: to make a public appearance or spar a few rounds demands less discipline.

By now Sullivan had given up the beer — and turned to spirits and champagne. And that was why a six months' interval came between the making of the match and the fight.

A famous conditioner, Bill Muldoon, was engaged; in later life Sullivan's face would blanch at the very mention of his name.

Even beer was forbidden as Muldoon laboured to rescue the real — or at least the former — Sullivan from his cage of flesh.

When the big day dawned, John L. looked like his old self — or rather like his young self.

It was just as well: the venue was Richburg, Mississippi — and not surprisingly the temperature soared to over a hundred.

This fight too had a strange setting: the ring was set up on a lawn before the mansion of a wealthy merchant.

And there amidst the trappings of Deep Southern civilisation the two Yankees fought a bitter and brutal battle.

It was rather like serving hooch in bone-china cups.

Charlie Mitchell was in Kilrain's corner: there was a contrast between the burning sun and the sleet and rain in France.

They fought for two hours and 15 minutes; then the official doctor warned Kilrain's seconds that their man was in extreme distress; they were not too surprised — and threw in the towel. And thus ended the last bare-knuckle fight for the world championship.

Sullivan again went on the exhibition trail. He claimed that he had no opponent worthy of a title fight — but this wasn't true.

The shadow of Peter Jackson hangs over John L's. record: the West Indian was a brilliant boxer — the champion, or perhaps someone associated with him, conveniently drew the colour bar.

Jackson's challenge ended when a young hopeful named James J. Corbett fought a draw with him over sixty-one rounds.

And so this rather unusual pugilist — he was a teller in a bank — became the new challenger.

And as every schoolboy knows, Sullivan and Corbett met in New Orleans in the first title fight in which gloves were worn.

It was rather like Mitchell and Sullivan all over again, only this time the boxer was much bigger than Charlie and the fighter was now 34.

The bout lasted for an hour and eighteen minutes: Corbett scientifically wore down the champion; Sullivan battled with mad bravery but was put down four times in the 21st round. On the fourth fall he failed to beat the count.

And thus the old regime ended and the new began.

Where the poor have only their pride . . .

Rosa is a professional singer in Mexico City. Sometimes when she goes out of doors, she wears a scarf across her mouth — this is when the air is especially foul.

And thus she hopes to escape from getting a sore throat, an affliction almost impossible to avoid in this conurbation which is at the head of the world's pollution league.

Mexico City has other hazards. Its drivers are not the world's craziest — that honour surely rests with Guadalajara — but they are strong contenders.

And although most of the city is almost as flat as Amsterdam, you will see very few pedal bicycles.

The reason is rather obvious: You don't compete against vastly superior forces.

Mexico City has its traffic laws but the gap between theory and practice is considerable. Indeed if you dutifully observe the laws, you will soon end up very dead.

And yet in over a month I have seen only one accident. The Mexican drivers have their own rules — and seemingly they work.

But the pedestrian who is given to taking chances is unlikely to have a long and happy life.

Minor hazards in Mexico's capital include the proximity of a few volcanic mountains and El Temblor.

The volcanoes are said to be extinct. A few thousand years ago the same belief existed about Mount Vesuvius — it struck back.

As this vast city grows and grows, people are building houses and shacks farther and farther up the slopes. It is a gamble but you must live somewhere.

El Temblor is the earthquake. Mexicans talk about it as if it is a person or an animal.

The city is built over a huge geological flaw. There were two shudders last Autumn. Thousands perished.

The official figure was given as 8,000 but that was only because Mexico was to be host for the World Cup finals in Soccer.

The locals put the figure at over 20,000. And even now, nine months later, not all the rubble has been cleared.

Immediately below my hotel window there is a shattered building. It would probably have been knocked down and the rubble cleared away if it had been nearer the centre of the city.

The Government have carried out a vast cosmetic job. The hordes of visiting journalists can go home happy in the belief that the reports of the Autumn earthquakes were grossly exaggerated.

The Press Centre is in a part of the city that escaped completely. It is also in a beautiful area of woods and fine buildings.

And the Press buses going to the games are carefully routed: The visitor sees a thriving city.

The Government have gone to ludicrous pains to show that all is well — as the people of shanty town Neza will tell you.

Neza is on the outskirts of Mexico City, if you can speak of outskirts in a jungle that is spreading and spreading.

Neza has a special claim to fame: It is deemed the worst slum in Latin America. And that honour was not easily won.

The authorities were not content to leave very well alone. A little while before the start of the football tournament they built high walls around Neza.

And the walls were decorative. An innocent passer-by might think he was seeing an old Spanish-type town.

This was bad enough but now the authorities have added injury to insult. They are demanding that the inhabitants of this earthly paradise pay for the walls.

The good people of Neza are not amused. This little story tells you much about Mexico.

The Government are striving to create a modern state but this is still a feudal country.

There is no apartheid in Mexico but there are vast gulfs between the classes.

The power elite are almost exclusively of Spanish descent. So are most of the middle class.

It is obvious that there has been little inter-marriage between European and native over the centuries.

It is obvious also that there are two tribes of people who for want of a better word I will call Indians.

One tribe are familiar from the movies that pretend to depict life in the Old West of America. They are the people who won every battle except the last.

And we can see them with their lean bodies and their high cheek-bones as they bite the dust in the final reel.

These Indians are well represented here in Mexico. You wouldn't have to search far to assemble an army to overwhelm Custer again at Little Big Horn.

There is another kind of Indian, if that is the word. These are small people with round smooth faces. They are built like cobs and are enormously strong. They do most of the heavy work.

And then there are the millions of Mexicans who have no work at all. How many, I do not know.

In Mexico City there are possibly five million with no regular employment. Many are country people who can no longer scrape a living from the land. Parts of Mexico haven't had rain in this decade.

And so they flock to the city. The older people have little alternative. Many of the younger people go to the United States on tourist permits and try to remain there.

And how do those people live who can no longer depend on the land? Only God knows. The tropical climate helps. Food is cheap by our standards — but when you are destitute, nothing is cheap.

The destitute ones in Mexico City live in tents and huts; those are lucky. Others sleep in doorways.

The authorities are doing all they can by building little pre-fabricated houses on waste ground — but their task is almost impossible.

The most amazing aspect of all this is that you will not see any beggars in Mexico City. In Ireland they say that a poor man has only his good name. In Mexico the poor will tell you that they have only their pride.

And so they do not beg. They sell something or perform some service such as polishing shoes or they entertain.

The variety of things offered by itinerant salesmen and women would give you a fair idea of infinity.

Some street vendors offer the obvious articles such as cigarettes and sweets and chewing gum — you will find such merchants in most big Latin towns.

But Mexico City is famous for the bewildering array of goods on sale in the streets.

You take the markets for granted — in Mexico City the displays of fruit and flowers and vegetables are almost stunning. But more fascinating are the people selling as they move along the sidewalk or dodge in and out of the traffic.

Thus you will see people selling all kinds of clothes, including beautiful blankets; more sell books and maps and papers; more sell spare parts for cars — there is no end to the variety.

The typical spare part is the windscreen wiper. The vendor carries them festooned around his head and body. On the first occasion I saw such a salesman I thought that he was a member of some new religion.

And always there are the little boys darting amidst the cars at the traffic junctions, with a can of liquid in one hand and a cloth in the other.

And they wipe the windscreens whether bidden or not. If they get no money, they just smile. The Mexican poor are well accustomed to getting "No" for an answer.

And of course, you cannot go for a hundred yards without being offered lottery tickets, as in Madrid.

And there are the street entertainers. You will meet them any time — they proliferate on Saturday nights.

Then parts of the city centre become pedestrian malls. Big groups perform — they play and dance and sing and juggle and tumble and otherwise entertain.

In Mexico you are never far from music. It is their expression and their anodyne. A little boy walks along the pavement in the morning and plays a whistle. His mother collects.

Rosa is one of the many beautiful singers who perform in the bars and cafés at night.

People from the outside world who have lived here for a long time tell me that they are still amazed by Mexico's vitality.

It seems to transcend the poverty — but of course, it doesn't. The worst of the poverty is invisible. And yet this is such a land of paradoxes that you cannot pass judgement.

Its capital, for instance, may be the dirtiest city in the world — but it is also the cleanest. The people cannot do much about diesel fumes but they keep their streets as clean as those of any town in Switzerland or Scandanivia.

And incidentally their favourite proverb is well known in my own part of the world "Tomorrow will be another day".

Evening Press, Tuesday, July 15th, 1986

Children — Mexico's greatest treasure

When Mexico's football madness was at its peak, their government told the world: "We cannot pay our debts — but we will pay when we can."

Next day a Mexican City daily carried an article headed : "Is football to be our foreign policy?"

Mexico's national debt is about a hundred billion dollars, so enormous a sum that it seems unreal.

But it is all too real and it isn't owed to a government who might be expected to write some of it off: it is owed to ten American banks who demand at least their interest.

Mexico's plight is all the more tragic because a few years ago this long-struggling state seemed to have burrowed a tunnel at the end of which there might be some light.

'Oil' was the magic word. And in the seventies Mexico borrowed billions to finance exploration and development. The black gold flowed.

But the bright gold didn't keep on flowing: at the beginning of this year a barrel of North Sea oil fetched 26 dollars — by mid-February it was down to a little over fifteen.

The fall in oil prices is a blessing to most countries: for a few especially Mexico and Venezuela it is a disaster.

Mexico's national debt now threatens the world's banking system: this year's interest amounts to about ten billion — it can be paid only by the expedient of more borrowing.

And Mexico has little hope of significant investment from outside.

The main deterrent is the government's bad odour: It may be as pure as the snow, driven or otherwise, but it has the name of being corrupt.

A first cousin of the president is said to be Mexico's leading drug baron; the chief of police is believed to be his closest competitor.

Another factor discourages would-be investors: Mexican bureaucracy is notorious.

You can see this even in such a simple transaction as cashing a traveller's cheque — at least it is simple in Western Europe.

The Mexicans revel in petty authority: The purchase of an air ticket can take up to twenty minutes — and then you may find that you are booked on a non-existent flight.

And so it is all too easy to understand why would-be investors put their money elsewhere.

Mexico's foreign debt isn't its only appaling problem: the growth of the population has gone into a spiral that must have consequences beyond imagining.

Officially it is now 83 million; it is probably nearer a hundred million — in Mexico City and in Guadalajara the babies are as plentiful as Milton's leaves that strew the groves in Vallombrosa.

And the poorer Mexico becomes, the more the population is likely to increase.

And there is little use in explaining to the poor why they shouldn't have big families.

The upper classes attribute the poor's fecundity to ignorance — that is patronising nonsense: Birth control is almost as old as man himself and has a part in the most ancient of folklores.

The Mexican poor will go on having big families because like the Irish travelling people they love children and look on them as about their only treasure in this world.

The Chicago of a century ago was described as 'a sprawling monster of a place, stuffed to bursting with extremes': Mexico City today would make it seem an extremely ordered village.

The part of it known as Neza is usually called a township — but its population is over four million.

It is deemed the worst slum in Latin America and it possibly is — but it has two redeeming features in the Mexican passion for cleanliness and their love of colour.

I saw only two graffiti in this urban jungle — 'The tourists do not come to see our hovels' and 'neither Castro nor Kissinger'.

The absence of graffiti isn't due to political intimidation but to the desire to keep their walls clean.

The love of colour is seen in the pots and pans and buckets of flowers with which they decorate their otherwise pathetic habitations.

Many families in Mexico City live in little tents on waste ground. The cooking is done in the open. There are no lavatories as such.

Most of these families have no regular income — you wonder how they survive.

The climate helps. The winters are mild. In Chicago in the 'black winter' of '93-'94 thousands fought for sleeping space in the corridors of public buildings.

And the Mexicans show infinite resource in the battle for existence.

I was told of a widow with a big family whom drought drove off the land: she came to Mexico City, set up a tent, bought a table, went to the market — and within two days of her arrival was selling home-made lemonade.

The Mexicans will sell anything: on my last morning in the capital I was approached by a young man festooned with trowels; I was tempted to buy one even though I doubt if I will ever again lay a stone upon a stone.

And I have never seen a market to compare with that beside The National Palace. It lines two sides of a long pedestrian street — the variety of things on offer seems endless.

And some of the stall-holders — or rather table-holders — are boys and girls of little more that ten; they would out-patter a tobacco auctioneer.

The National Palace is a huge and beautiful building: it makes its Buckingham counterpart look like an ageing Portakabin.

It forms one side of a great plaza. I sat there about six o'clock on my last evening in Mexico — and was attacked by a bewildering array of sensations.

The most abiding was deep loneliness, the loneliness that comes from feeling that you are looking at a world of which you can never be part and which you cannot even begin to understand.

There was also the intuition that the Mexicans have retained much that we in Ireland have lost.

The plaza was thronged with homegoers — and all over the place you could see couples dallying for a few embraces before parting for a bus or for the underground.

And should a sex therapist ever set up business in Mexico, he might soon find himself begging for bread.

'Sex' is a distasteful word — but hard to avoid. The Mexicans are possibly the world's last romantics — almost all their songs are about the joys and woes of love.

Rosa is a chanteuse who sings most nights in a small hotel near the city centre. Like the heroine of Theodore Dreiser's great novel 'Carrie' she is a country girl who came to seek fame in the big city.

She is now about thirty and fearful that she will never be more that a café singer.

She lives on her own and misses her people far away in the north near Monterey. Her personal loneliness infuses songs that are intrinsically sad.

She performs twice a night in stints of about an hour each; the intervals are filled by two remarkable men who play guitar and sing together.

They have a trade name but nobody seems to know their real names or indeed anything about them, except that they are consummate professionals.

Both are in their sixties. One is smallish — the other is very small. They wear sober three-piece suits and are dandies as only Latins can be.

They are very friendly but will never take a drink: when their night's work is over, they put away their guitars, pick up their little rolled umbrellas — and disappear.

They sing in a way I have never encountered before. 'Plaintive' comes to mind — but it is a hopelessly inadequate word.

Mexico abounds in romantic singers but these two little men in their sober suits seem to be expressing a harder world than even their troubled country knows today.

Sometimes late at night the head barman, Arturo, joins them. He is low-sized and built like a badger. You would guess that he is a descendant of the forest Indians — and he sings in a way that makes even the two little men seem close to European music.

And invariably he ends with a stream of what are like animal cries — you could be listening to a fox in the lonely nights of Spring.

And between Rosa's husky sadness and the lost-world singing of the two little men and the forest voice of Arturo you are left at once desolated and exhilarated.

And you are left too with a tremendous sense of layers of civilisation intermingling and of an enormously complex country which nevertheless seems to be home to a nation.

What the Mexicans of all classes seem to have in common is a passion for music and a love of colour — nowhere else have I seen builders' helmets in a wide array of hues.

And on Saturday nights when all heaven breaks loose in the streets and the bars and the cafés, the national debt hardly seems to matter.

But even on those festive occasions the reality is seldom far away: it is sad to see fine-looking old men, once members of a proud peasantry, having to get by selling chewing-gum.

And the hysteria that took over Mexico for most of June was rooted in abysmal poverty. It came from the bottom, from the have-nots.

They hungered and thirsted for a sign that the gods hadn't entirely forgotten them. And they knew that they had their best team ever.

And three faces out of that tumultuous time I will always remember.

One was in a crowd of young people in Guadalajara who were drumming on the roofs of cars at a traffic junction and chanting 'Mexico, Mexico, ra-ra-ra.'

He was a tall lad of about sixteen. Obviously he had been in an accident: his face was scarred—and his upper body was twisted.

Mexico were still very much in contention for the cup—and there was a madness in that boy's eyes, as if he had a vision of a victory which would transform him and his people.

And there was the boy with no arms whom I saw playing a mouth-organ and a drum outside Azteca Stadium on the day of the final—his vitality symbolised Mexico.

And there was another boy: I saw him on my third night in Mexico City; he was playing a button-accordion in the Calles Londres, at the heart of the Saturday night madness.

He was about sixteen. His face was pure Indian and his eyes were as sad as his music—the tune was like that played by the blind man near the end of 'Amarcord.'

A little girl of about fourteen held out the can for the few coins that came her way.

A few yards away a child who might have been two was dancing to unheard music—all he wore was a shirt that came as far as his navel.

After a while it dawned on me that they were father and mother and son.

A man obsessed with the essence of things . . .

Derek Hill would not be instantly recognised if he walked into Mulligan's or any other of Dublin's semi-bohemian pubs — or even if he ran in; despite this flaw in his social armour, he is one of the ablest painters working in Ireland today.

He was born into a wealthy family and enjoys (or otherwise) a private income; he never pretended to live in a garret on onions and brown bread and rough red wine; if he has manned barricades, they were spiritual — he is far from being a typical painter, real or alleged.

Worse is to come: he admires Augustus John's portraits; that solecism is in itself enough to damn him in the eyes and ears of those who fancy themselves as the pioneers destined to take art across new horizons.

About two years ago the BBC put out a fine documentary on his work; I met very few who watched it or even heard about it.

He has had seven one-man shows in this country between 1960 and 1982 — in Belfast and in Dublin and at The Wexford Festival — and he is represented in our National Gallery and in The Hugh Lane Gallery, yet he remains an unsubstantial figure to the generality of Irish art-lovers.

His origin is possibly part of the reason: Derek Hill is as English as oast-towers and high tea and perry, not to mention the maypole and pale ale and cottage cheese.

He was born in 1916 in the richest part of our neighbouring island, the south-east.

His father had perceived that the English climate would indefinitely create a need for fuel and raiment — and so he sold coal and cloth. He prospered.

For good measure he captained Hampshire at cricket; his team-mates included W. G. Grace.

His other passions were fishing and shooting. He hated music so intensely that his wife, an accomplished violinist, never again played after she married.

A. J. L. Hill seems an unlikely father for an artist but seemingly he and young Derek got on well.

When Hill junior announced that he intended to be a painter, the old man made no objection; indeed he gave him a generous annual allowance.

It was a remarkable display of understanding; the budding painter was only sixteen.

And for most of his young life it was roses all the way.

In Paris he fell fair of a remarkable man: Edward Molyneaux had started his working life as a newsboy in London; he became the most celebrated couturier of his day.

He was also an art collector; he acquired works by Manet, Corot, Monet, and Van Gogh at a time when Impressionism was distinctly unfashionable.

He persuaded the 22-year-old Hill to give up stage designing and go flat out as a painter.

He put a chauffeur-driven car at his disposal to take him wherever he wished to go in Paris and its environs.

And, more important, he gave him the run of his apartment; there he could study The Impressionists in a way not possible in a public gallery.

The war disrupted this idyll; Derek Hill came home and registered as a conscientious objector.

It was a brave but surprising decision; he had experienced the jackboot in the flesh when The Nazis took over Austria.

There could be no doubt about his honesty; the army employ painters as chroniclers rather than soldiers.

And, unlike some famous poets of the day, he didn't flee to America; he worked on the land — and helped to man an anti-aircraft battery.

Eventually peace was declared.

Derek Hill came to Ireland in 1946 and painted in Galway and in Achill.

It is said that the longing for travel is akin to the urge to write — and to paint; Oliver Goldsmith would agree.

Travel was difficult in his day; Derek Hill's peregrinations make it seem that poor Noll had only been down the road to borrow a cup of sugar for his mother.

In 1954 his wanderings came to an end — more or less; he bought a house in Donegal and in the meantime has lived there more than anywhere else.

Eventually Tory Island became the principal matrix of his landscapes.

Tory is a desolate rock, partly covered by a thin skin of soil; its inhabitants could hardly exist without Government assistance and remittances from emigrants.

And yet this God-forsaken (and eventually it will probably be man-forsaken) spot on the map seems to have inspired Derek Hill.

His long-time friend, Grey Gowrie, (better known as Lord Gowrie, chairman of Sotheby's) seeks for the answer in a book published recently.

In the introduction he refers to his subject as "the best painter of the Irish Landscape since Jack Yeats" — fighting words, you will agree.

And he passes a wicked judgement on Paul Henry: "He always seems to have one eye on how his paintings will reproduce on the bedroom walls of exiles."

That, of course is nonsense: Henry's vision may have been a trifle sentimental — but it was his own.

He is deemed an unmentionable by the vanguard of painters and critics today because the scenes he depicted are recognisable.

If you applied the same criterion to portraits, it might seem to be invalid.

Derek Hill is a celebrated painter of portraits — and not all are of 'establishment' people.

He depicted Freddie Mills, the great boxer — he refers to him as 'a good and kind man'.

Most of his portraits were not commissioned: he painted people because he liked them: there was no obligation to buy.

And he tells Grey Gowrie in this frank book that painting a portrait was his way of getting close to someone.

Derek Hill never married, although thrice on the brink; seemingly he never had what is commonly called 'an affair' — and finds it very difficult to have a solid relationship with someone.

Grey Gowrie obviously suspects that this sense of lone-ness is the key to his fascination with Tory.

He has depicted some of its people but his principal work there has been a series of attempts to 'capture' the essence of the island's physical self.

And that demands something far more difficult than what Paul Henry did in the west of Ireland.

Derek Hill is obsessed with the essence of things; the north-west with its ever-changing light has proved to be his severest trial.

His land-and-sea-and-sky scapes are abstract in the sense that local people would not readily recognise 'the originals'.

That, no doubt, was why an islander, Jimmy Dixon, said to him one day: "I could do better than that".

Derek Hill gave him the materials. Thus started the famous 'school' of Tory's 'primitive' painters.

Grey Gowrie accuses him of not allowing the human aspects of the island to speak; his explanation is that what he must express is the harshness and the austerity.

You will not be surprised to hear that The Burren is high among his favourite places.

He is mistaken, however, when he states that no one but himself has painted it; for some years now it has been the Mont Saint Victoire to a friend of mine.

Tory — like The Burren — seems a vast remove from art's capitals but the man who lived for years in Paris and in Rome is obviously at home there.

Evening Press, Tuesday, November 3rd, 1987

Van Gogh, unquiet spirit

It may seem ridiculous to compare Lester Piggott and Vincent Van Gogh: each, however, was consumed by a passion that led to sacrifices so enormous that 'normal' people can hardly comprehend.

The obvious difference is that Piggott earned vast sums from his craft — Van Gogh earned next to nothing from his art.

The great jockey also accumulated an abundance of objective evidence to assuage the demons of self-doubt; the great painter died destitute of money and almost destitute of acclaim.

Van Gogh's failure to achieve general recognition is mystifying to us; his greatness seems to leap from the canvas. Why did his own generation not see it?

Part of the answer lies in the power enjoyed by the art critics in nineteenth century France; they could make or break in the manner of New York's drama critics today.

Professional critics are notoriously conservative; they do not like to be disturbed by originality.

Novelty they can take and even applaud — but they fear originality as much as a child dreads the darkness of the night.

There was also the factor known as fashion; it can enslave all but the boldest of spirits, — and Van Gogh was woefully outside what was deemed good painting in the France of his day.

While in art school in Paris, he was deemed a buffoon by his classmates, not one of whom is now remembered.

Ironically, he was to become the darling of all kinds of students a generation after his death.

Prints of his 'Sunflowers' were as obligatory in bedsitters as photos of Che Guevara became a generation later.

'Sunflowers' was a kind of mantra, a star of hope for young strugglers, a symbol of the promised land.

144

It isn't a great painting but it glows with freshness, the quality perhaps that most alarmed the critics.

And it is a quality which Van Gogh shares with those contemporaries of his who were dubbed the Impressionist school.

The word 'school' is misleading; most of those included under that umbrella worked alone; what they had in common came from the spirit of the times rather than from regular communion.

Theodore Dreiser in his autobiography tells about his origins as a writer: he grew up with Chicago; the vitality of the era forced him into expression.

Paris was an old city by the last generation of the nineteenth century but it vibrated with new life; the Franco-Prussian war was among the causes.

For France it was a terrible catharsis: the appaling incompetence of the government was laid bare — and the division of the classes was starkly exposed.

After their betters had shamefully capitulated, the common people of Paris resisted in a way that pushed out the frontiers of heroism; they redeemed France's honour; their reward was to have the government's guns turned on them.

It would be a romantic illusion to think that writers and painters and sculptors and composers were among those who manned the barricades; a few were; most shunned the struggle — artists tend to have an extreme sense of self-preservation.

They could not, however, but be affected by the new climate; the cleansing effected by the war deepened the concept of democracy.

Edouard Manet's masterpieces were produced after 1870; it was hardly a coincidence that both depict humble life.

In much of Van Gogh's early work the missionary threatens to take over from the painter.

In his apprentice years he lived for some time among the poor of Belgium and Holland; he produced a series of drawings which are powerful but oppressive.

You see men digging and women carrying burdens and people waiting their turn in the public soup-kitchen; all seem weighed down with the woes of the world.

Van Gogh was utterly free from the slightest taint of humour; even though these early drawings are intensely sympathetic, they had little understanding of people.

His miners and fishermen and peasants and over-worked women seem to be darkened by the painter's own depression — they lack totality.

Those people looked for little from life; without a little yeast of spirit they could hardly have survived — you will not find it in Van Gogh's drawings.

And yet that desolate period of his life led to a masterpiece, 'The Potato Eaters', a work at which he laboured off and on for four years.

The people who inhabit that most ambitious of all his paintings are sad too — but the picture glows with a kind of sombre luminosity.

He produced far more drawings that paintings during his stay in The Low Countries, probably because he couldn't afford canvas and paint.

Later in Provence he was so poor that he sometimes worked with quills and reeds.

In between he had lived for a while in Paris. There he produced another masterpiece; he christened it merely 'Montmartre'.

That now famous quarter was then more or less a rural village; Van Gogh's picture is an act of homage to it.

It was painted from the terrace of the Moulin de la Galette; below lies the great city but Van Gogh chose not to show it — what we see is a kind of infinity in blue.

It is an amazing picture: there seems to be little in it — but it is endlessly evocative.

The Van Gogh we know from 'Sunflowers' belongs to his time in the south of France.

It was his best period: there he produced an abundance of splendid paintings — the most famous is probably his portrait of the old postman who was one of his few friends in Arles.

That marvellously fertile period ended in a nervous breakdown from which he never fully recovered.

His beloved and endlessly kind brother, Theo, an art dealer in Paris, got him into a mental hospital about 40 miles north of the city.

The head doctor there was a man of great understanding. And as an amateur painter he took a special interest in Van Gogh — but Vincent was now beyond humanity's reach.

The work he produced in that last period of his life showed a deranged mind.

One day while trying to paint in the fields, he shot himself in the chest; where he got the revolver was never found out — he had the cunning of the insane.

He didn't commit suicide: his action was the proverbial cry for help; the wound would not have proved fatal if he hadn't so woefully neglected his health.

On the day of his death he appeared on the road to recovery: his mind was clear; he seemed content, almost happy — but his mortal life was over. He was 37.

He had been a full-time artist for only eight years — he produced over eight hundred paintings and innumerable drawings.

The story of Vincent Van Gogh was been outrageously romanticised, not least in that appaling film 'Lust For Life'.

In truth, he had a terrible existence: it was compounded of poverty and loneliness and frustration, relieved by fleeting moments of ecstasy.

Vincent was a man of superb intelligence; you can see that from his letters — but he was too intense for what we call the world.

He failed at several professions before he hurled himself into painting; he might never have been more than an amateur artist if he had succeeded in 'normal life'.

There was nothing of the revolutionary in Van Gogh; he was as middle-class as you would expect from the son of a Dutch pastor.

He longed for a steady income and a home and a wife and family — but it was not to be.

Nor did he see himself as a revolutionary artist; he adored the old masters and was steeped in their work.

In a sense, his mission was a cause without a rebel.

And yet we cannot help thinking of him as a revolutionary.

'Sunflowers' is like an explosion of light and it is a happy painting.

It evokes images of the young rebel artist working in a milieu that inspired him; alas, the people of France are as conservative as any in the world; it is no coincidence that the word 'bourgeois' comes from their language.

It took a long time for Vincent's work to be appreciated; as a contemporary painter said about himself "the wedding-feast of the world passed me by".

A voice from the Deep South

"Jewel and I come up from the field, following the path in single file. Although I am fifteen feet ahead of him, anyone watching us from the cotton-house can see Jewel's frayed and broken straw hat a full head above my own."

In that opening paragraph from 'As I Lay Dying' you can see much of what is good and what is flawed in William Faulkner.

Above all, the sage of Mississippi was bold, sometimes too bold as in that "fifteen feet ahead of him" — almost any other writer would have had recourse to 'about' or some such word.

Of course, you immediately forgive him: his way of expressing the difference in the heights is brilliant.

Do not excuse Faulkner because this is first-person speech; it was a device he often adopted. And it challenges the belief that in writing you must always strive for economy; the difference could have been conveyed in a few words — but conveying isn't always sufficient.

That question used to torment Gerard Manley Hopkins; he put it most starkly thus — "How do you make poetry out of a man putting on his clothes?"

Thomas Wolfe didn't worry about it: the giant from the mountains of Carolina scorned all theories about writing; energy and intuition were his launching pads.

It is hardly surprising that he was the writer whom Faulkner admired most. Both had Southern rhetoric in their blood: Faulkner's description of his writing — "A life's work in the agony and sweat of the human spirit" — could have been spoken by Wolfe.

It is, of course, at once sentimental and pretentious — and yet acceptable; Faulkner made his own values.

Of all the modern writers in English he is perhaps the most difficult: 'Finnegan's Wake' needs only patience; Faulkner demands almost a new vision of the world.

There is a comforting belief that mankind is the same from China to Peru but it isn't true.

Robert Frost put it simply: "How can you write the great Russian novel in New England while life goes on so unterribly?"

Faulkner might have said: "How can you write with economy and lucidity in Mississippi while life there is so diffuse and turgid?"

Faulkner's work is easy reading only in patches: they are like little clearings in a tangled forest or small pools in a struggling river.

Those intervals are so good that they almost persuade you to see the generality of the work as penetrable — yet I know great lovers of literature who have given up what they see as an unequal struggle. And this raises a disturbing question. Is there an absolute criterion? A subsidiary question is: "Should you know Faulkner's world to understand his work?"

Reluctantly I must conclude that the answer to the second question is 'Yes'.

To some degree this is true of many writers: there are, for example, metaphors in Patrick Kavanagh's poetry which impinge more forcefully on those who knew life on an Irish small farm fifty years ago.

They, however, do not take from the body of his poetry: generally it is instantly appreciable; the generality of Faulkner's work is not.

Why is his world so different from ours? The question evokes an ancient metaphor: there are times when you find it almost impossible to reach a patch of firm ground whence you may begin to drain a marsh.

What Faulkner and Wolfe have in common is their enormous energy, so driving them that sometimes they are almost inchoate.

Their world has rightly been called The Deep South.

It is an unknowable world — so vast are its mountains and swamps and forests.

It has what we in Ireland long for — an outback.

Our outback is elsewhere; for most of us it is in the other island — and especially in London.

You realise this when you meet your neighbours home on holidays.

After a few drinks they intone such names as Shepherd's Bush and Paddington Green and Notting Hill Gate as if they were places of high romance.

It may seem pathetic but it is easy to understand; you need a world in which to lose yourself so that you can find yourself.

Faulkner's outback was all around him. He expresses an aspect of it in 'Go Down Moses'.

"He was sixteen. For six years now he had been a man's hunter. For six years now he had heard the best of all talking.

"It was of the wilderness, the big woods, bigger and older than any recorded document — of white man fatuous enough to believe he had bought any fragment of it, of Indian ruthless enough to pretend it had been his to convey."

That is one aspect of Faulkner's world; with it goes a human complexity; The South was as much a melting pot as New York or Chicago.

In a way it wasn't a melting pot at all; the elements didn't coalesce.

There are communities in The South who have no official existence; they manage to evade the census takers.

149

They live in the vastness of the mountains, a vastness marvellously captured by James Agee.

They have their own reasons for preferring anonymity; fear of conscription is not the least.

Sociologists include them among 'the poor whites,' the people who figure large in Faulkner's work.

They are illiterate and pathetically ignorant but are possessed of a fierce life-force.

In their own way they see themselves as 'upwardly mobile'; in contrast are the decaying aristocracy, the families who built their empires on the black slaves.

And, of course, through almost all the literature of The South runs the memory of The Civil War.

In Southern myth it has become the war between the pragmatists and the romantics, the Yankee versus the gentleman.

'Gone With The Wind' is not a great novel nor was it made into a great film but it is easy to understand why both novel and film became so enormously popular.

Wolfe in a strange rambling letter to his publisher, Maxwell Perkins, expresses what is suspiciously like Southern romanticism at its most mawkish.

The theme of that epistle could be summed up as 'You can never win'.

You won't find many winners in Faulkner's work.

One of his 'fulfilled' people is a man who has been sent to prison for a long stretch; he is happy with the thought that when he gets out, he will kill his condemnator.

Faulkner's love of the interior monologue gave him great scope to exploit legitimately the common language of The South.

The device was far from original but few took it as far as he.

And it is a marvellous way of capturing character.

Listen to Cora in 'As I Lay Dying'.

"So I saved the eggs and baked yesterday. The cakes turned out right well. We depend a lot on our chickens. They are good layers, what few we have left after the possums and such. Snakes, too, in the summer. A snake will break up a hen-house quicker than anything."

Sometimes you wonder how such poor uneducated people can be so articulate — but it is a fact of life.

And not the least of Faulkner's virtues is that he perceived drama in such seemingly simple lives.

He might have been exemplifying Thomas Hardy: "At the graveside of even the humblest man you see his life as dramatic".

And I laugh when I read or hear that the novel is dead.

It is about as dead as dreaming and loving and drinking and eating and all the other activities which make up what we call 'life'.

Sunday World

A classic in the true sense of the word

The good women of Brabant milked the cows on the morning of the Battle of Waterloo and they washed the children and fed the pigs and the calves and the fowl. Life went on.

Then they kept an eye on those hens that liked to lay their eggs in secret places, no easy task I can tell you. The men did not go into the fields though it was mid-June: for a start, it was raining — and for a finish, it wasn't a day to be out of doors.

A friend who was in the Dieppe landing told me that the birds sang during lulls in the fighting. Of course they did but they weren't really singing — they were expressing their alarm.

What is more remarkable is that racing went on in our neighbouring island throughout World War Two, if only in a greatly diminished form.

Newmarket was and is the capital of British racing; no part of England was more vulnerable to aerial attack. The people of Britain were bracing themselves for invasion in the Summer of 1940 but the Classics were staged.

Newmarket was the venue. The Derby was held outside Epsom for the first time in almost 150 years — The Downs were occupied by the Army.

In the February of 1941 a bomb fell on The White Hart in the middle of Newmarket: 27 people lost their lives and about 300 were injured — but The Derby wasn't affected.

Indeed so great was the crowd that many had only glimpses of the race — and seemingly despite the acute scarcity of petrol, there was traffic congestion.

Of course not everybody approved: some of the Labour members in The House Of Commons, notably Herbert Morrison, saw all this as another example of class division. He was hopelessly wrong: racing, at least in these islands, is democratic in its own peculiar way.

Men and women who have never owned as much as a square perch of land can look on certain horses as their own. You can be sure that the good old working class was well represented at those wartime classics at Newmarket.

Herbert Morrison deemed it criminal that racing should go on while people were fighting for civilisation; Winston Churchill argued that racing was part of that civilisation.

Those of you who are younger may think that wartime racing was necessarily of modest quality — it wasn't that way at all.

Watling Street, winner of the 1942 Derby, was deemed a colt of especial class — but Sun Chariot, winner of The Oaks, made him look like a pretender in the St. Leger.

Indeed a question that perplexed all racingdom was her connections' failure to run her in The Derby. Gordon Richards, her pilot, used to say that he couldn't imagine her equal.

There is a tendency to look down on wartime Derby winners because the race was staged at Newmarket. There are aficionados who would put up a counter argument — luck played less a part at Newmarket because it is a fairer track than Epsom.

They have a case: the Epsom track resembles a horseshoe that went wrong on the anvil; the Newmarket track seems to have been designed to eliminate chance.

Of course it can be argued that Epsom is the better test because it demands so much from horse and jockey.

The race blasts off uphill for about two furlongs: then for about two more furlongs the track is downhill until the turning starts for Tattenham Corner. Of course it isn't a corner at all: it is a great sweeping curve that is downhill all the way.

It is still downhill as it turns into the straight: then after about a furlong and a half, the track rises — the last two furlongs are uphill.

You might think that only a jockey used to its rises and falls and curves could win the Epsom Derby but it isn't that way: Lester Piggott won the big race when most of his contemporaries were still at school.

Willie Shoemaker almost won it on his first visit to Epsom. I could sing out a list of jockeys, including Neville Sellwood and Yves Saint-Martin and T. P. Glennon, whose lack of experience didn't prevent them from winning.

A good jockey needs only to walk a course to fix its map on his mind's screen.

The last wartime Derby was a famous occasion; it attracted the biggest crowd seen at Newmarket before or since. There were queues at dawn in Liverpool Street station: many people cycled the odd 60 miles from London.

152

On that June day 56 years ago i was working in the bog in the parish of Knocknagoshel but in spirit I was in the heartland of racing in our neighbouring island.

I had a vested interest: Dante was my favourite horse and still is — perhaps I was attracted to him by his name. There is a mystical belief that great horses have great names — Dante exemplified it.

I couldn't see him being beaten in The Derby, even though he had lost in The Guineas. His connections blamed that defeat on the fact that he was blind in one eye.

This didn't seem to bother him in The Derby: he won in such style that many aficionados put him among the immortals. I had a nice bet on him. I was confident that he had won even though I didn't find out the result until late that night.

Many years later I was pleased to discover that Phil Bull, the legendary racing guru and gambler, got his "foundation" money from Dante.

I have said it often in pub and in print that if you can afford it, you should go to The Derby at least once. In this context it resembles St. Paul's Cathedral and The Castle Rock in Edinburgh and The Crown Bar in Belfast.

When you stand on Epsom Downs in the company of perhaps 200,000 people, your troubles fall from you. It is a kind of "other-world" experience: you can believe that Fred Archer's ghost rides the track on his favourite horse, Ormonde, on the night before The Derby.

If Lester Piggott ever dies, you can be sure that his ghost too will go down the hill and around Tattenham Corner on his favourite horse. Will it be Sir Ivor or Nijinsky or The Minstrel?

It could be Empery, a horse that is almost forgotten. Lester won The Derby with him in 1976. It was a sweet victory because he wasn't fancied. He started at 10/1. The price would have been twice as much if Lester hadn't been on board.

Let us return to Herbert Morrison: he was a good man and left his stamp on the world but he was wrong about the place of sport in life.

Racing was a part of the war effort: it brightened people's lives.

They had few other anodynes, except watery beer and a half ounce of tea a week per person and a sinister concoction called spam.

This was defined in Chambers' Dictionary as "a luncheon meat made from pork, spices etc." I like the et cetera.

———————

A battle with the scales for Piggott and Archer

It would be a strange thing if I hadn't an icon from the world of Racing. It is the universal sport. It appeals to all classes and ages and genders.

"All men and women are equal above the turf or beneath it" may not be totally true but there is a kernel of truth in it. Racing transcends divisions.

It used to be said one time that crowds would wait for half-an-hour or so to see Prince Edward, the future King — but might wait half-a-day to catch a glimpse of Fred Archer.

The legendary jockey was born in Cheltenham in 1859. His father was a jump jockey, one of the many brave but obscure men who underpinned the game without ever achieving fame or fortune.

Young Archer was apprenticed at thirteen to Mat Dawson, the most famous trainer of the day. He was with him only a few weeks when the great man said: "I have got a wonderful child who will achieve wonderful things."

Never was a prophecy more justified. Such was the young man's success that his name became a byword. When you said "Archer's up", it meant that everything was going your way.

It would be impossible for us today to understand his fame. Combine the adulation that surrounded Elvis Presley and the Beatles and Georgie Best and you have only a glimmer of the reality.

Racing and Boxing were the only major sports of the day. Organised Football was only just starting. Television and Radio weren't even pipe dreams. Hollywood had yet to come and transform the world.

Archer fulfilled the people's need to worship. Women of course adored him, even though for most of his career he struggled with the scales to such a degree that he looked like a walking ghost.

He could have married into Royalty but he chose one of his own — a girl who couldn't have had a more English name: Nellie Rose was a niece of Mat Dawson, his original master and most fervent admirer.

They married in 1883. A special train brought them from Newmarket to spend their honeymoon in London. That year marked the pinnacle of Archer's career.

By the second half of 1886 the gods had ceased to smile on him. Nellie's first child had died in infancy. This added to the depression caused by his weight trouble.

He was defying medical opinion. He was almost six feet but for some races he got down to eight stone three pounds. There were occasions when he was known to fast for three days.

His idea of lunch was a sardine and a glass of Champagne. To survive and keep on riding he depended on a purgative known as Archer's Mixture, concocted by a Newmarket pharmacist.

Their second child survived but Nellie died shortly afterwards. Archer soon followed her. On a dark November evening in his gloomy mansion in Newmarket he took his own life. He was 29.

In legend he has come down as the greatest jockey of all time. His partnership with Ormonde is deemed the peerless achievement of man and horse together. They won The Triple Crown — the 2,000 Guineas, The Derby, and the St. Leger.

Ironically, many of the aficionados of Archer's day didn't judge him the best of his generation, let alone of all time. They gave the accolade to a man whose name isn't remembered today except by historians of the turf.

Many of their fellow jockeys would say that George Fordham had better hands and better vision and was a better judge of pace. He had one obvious advantage over Archer — he had no weight problems.

He could live a normal life and always ride at full strength. He loved food and drink and had a passion for Port. Once when told to give a bottle of Port to a horse, he administered it to himself.

He was also deemed the cleverest rider of his day; Archer was always uneasy unless he had him in his sights. He used to say: "I am never happy when I can't see him and know what he is up to."

Fordham was often up to something: he was a genius at pretending to have fallen out of contention and then biding his time to come again.

In another respect he was a better jockey than Archer: he was gentle with horses and sometimes, especially with fillies, he discarded the whip and the spurs.

In this he resembled two great jockeys of later years — Steve Donoghue and Garnet Bougore. Archer was ruthless with horses. Fordham was a much nicer man but he is almost forgotten whereas Archer is immortal.

The resemblances between him and Lester Piggott are obvious. Both were sons of jump jockeys. Piggott was born into a better-off family but he had his own difficulties: he was hard of hearing and he had a speech defect.

Both throughout their careers had to battle endlessly with their weight. Once when Piggott took a short Winter break and ate normally, he went up to eleven and a half stone.

To ride at 8 stone 7 pounds, as he did for most of his career, required the self sacrifice that would make a Trappist monk seem like a playboy.

If he was hard on horses, like Archer, he was harder on himself. They had another thing in common: they had no time for comradeship — they thought nothing of asking for a mount even though a friend had already got the job.

Archer's life ended sadly. The gods spared Piggott the ultimate tragedy but played a bizarre trick on him: when he should have been in safe harbour, secure for life, he was reduced to pauperdom.

He was the unwitting victim of a tax fraud, organised by two leading trainers and two wealthy owners. Archer had declined many a Royal invitation; Piggott had no choice but to spend a term as a guest of Her Majesty.

He wouldn't have been imprisoned in "a poor backward country" like Ireland: the British are different; on a snowswept morning in the January of 1649 their King, Charles The First, paid with his head on a scaffold in Whitehall.

Lester Piggott made a famous comeback and it couldn't have been more fitting that he did so in this country.

George Fordham achieved fulfilment, something that seemed beyond Fred Archer and Lester Piggott — they were driven by demons that most people cannot understand.

Evening Press, Tuesday, April 8th, 1980

O'Casey — The First Careful Rapture

Some people will tell you that Sean O'Casey's volumes of autobiography are more important than his plays — and perhaps they are right.

All I know is that I started to read them all — and came down even before I reached Becher's.

His prose is often abominable — and his sentiments sometimes astonishingly mean.

And he was not untypically Irish in his believing that he had a licence to flay everyone but should be immune from criticism himself.

One might forgive him that piece of Double Think if only he showed more respect for his readers.

Communication is a two-fold activity — a fact that O'Casey's self-indulgence often caused him to forget.

The Hollywood producer who asked him write the screen-play for 'Look Homeward Angel' was no fool.

He sensed an affinity that is far from obvious: Thomas Wolfe, like O'Casey, was so excited with what he had to say that he did not always succeed in saying it.

Wolfe had a dedicated editor, Maxwell Perkins, who distilled much of his outpouring; O'Casey was not a man to submit to such a process — and could mistake correction for censorship.

It was hard to blame him: before he achieved power, he had suffered so much from the expurgators that in later life he understandably declared himself infallible.

But the theatre had one salutary effect on him: his words had to be spoken — and understood at the first go.

And in the three plays on which his popular fame uneasily rests, the communication is both simple and deep.

Academic critics often talk about a writer communicating at different levels — it is a concept founded in snobbery.

And whatever failings O'Casey had, he did not write with one eye on the intellectuals and the other on the class from which he had more or less emerged.

Ayamonn in 'Red Roses For Me' expresses O'Casey's attitude. He is talking about Shakespeare: "They think he's beyond them, while all the time he's part of the kingdom of heaven in the nature of everyman; before I'm done, I'll have him drinking in the pubs with them."

Ayamonn, of course, is as near to the young O'Casey as Paul Morel in 'Sons And Lovers' is to the young D.H. Lawrence.

And even if in the play he is a little bit romanticised, he nevertheless indicates the folly of much that has been written about the hardship O'Casey endured in the years before fame smote him.

Of course he was poor and of course he worked at laborious and ill-paid jobs. In that he was one of millions in these islands.

But he was different: he had the magic of mind. He expresses that truth through Ayamonn when his mother advises him to rest a bit and not to be trying to be a full-time workman and full-time student.

"You're overdoing it. Less than two hours sleep today and a long night's work before you. Sketchin', readin', makin' songs, and learnin' Shakespeare . . ." Ayamonn answers: "They are all lovely, and my life needs them all."

And it is true that from childhood O'Casey was threatened with blindness — a threat that eventually materialised, more or less. But it made his awareness all the sharper.

And being less than affluent is not always bad for a writer. If you wanted an account of a wedding-feast, you might be well advised to skip the guests and ask some poor devil who had been watching it from the outside.

O'Casey's supposed handicaps were really spurs — and the common verdict that he was self-educated is just as silly as the talk about his hardships.

The same adjective is often applied to Thomas Hardy too — why, only heaven knows.

Hardy didn't attend a university — but he got a long and thorough academic education.

O'Casey seemingly was never at school at all — but he lived in a great age of self-improvement: he was in part a Victorian.

And the learning quarried out in hard-won intervals of leisure is not forgotten. O'Casey had a great hunger — the kind that academics often lack: the proverb about taking a horse to the water has a wider application.

And he was a Protestant — and heir to the Bible and a great hymnology. The imprint is clear.

The other obvious influence is Shakespeare, especially in "The Silver Tassie".

The language in that play is highly improbable — but no more so than in 'Hamlet' or 'Macbeth'.

It may seem pretentious to make a comparison between Elizabethan London and Dublin at the start of this century — but there is an affinity.

Shakespeare's London was small and foul and pestilential — but it was inhabited by a people who were suddenly becoming aware of their identity and who were intoxicated by it.

O'Casey grew up in a Dublin of appalling poverty — but of intense ferment. He was part of two streams that never fully coalesced — the nationalist and the socialist.

And to say that O'Casey — and many another — was dragged in different ways by two wild horses is not to stretch a metaphor.

O'Casey rejected nationalism — his two big 'Dublin' plays are sufficient proof.

He knew well that those who appear to succeed by the gun are not likely to throw away their arms and yield to democracy — 'freedom' can take strange shapes.

The dragging away of Johnny Boyle to be 'executed' by his old comrades is a microcosm of the Civil War.

And, as O'Casey well knew, one of the early sessions of the Sinn Fein parliament produced a proposal to reduce the old-age pension.

He had another reason for standing aside from the wave of nationalism — he was a Protestant.

And, as Canon Sheehan had forecast, the time came when the complex movement that had won a measure of independence would be grossly simplified.

The non-conformists had been its principal begetters — and Protestant involvement had continued. But when victory (of a kind) came, it was soon advertised as a Catholic triumph.

The Sam Maguire Cup is a symbol of the metamorphosis. It is named after a man who was a Protestant and a Fenian. But the Association for which he did so much is now militantly Catholic.

It is said that Sam Maguire died of a broken heart. And well he might. The new Ireland for which he had sacrificed so much looked on him and his fellow-Protestants as enemies in the camp.

Many of them went away. And those who remained had the privilege of being regularly reassured that they were, at best, second-class citizens.

O'Casey went away — and did not come back. Many reasons have been given for that exile — the most probable was that he could not say 'Yes' to the new State.

And he took with him a great bitterness, like that of a man who has played for the team through thin and thinner but finds that there is no place for him in the cup final.

Out of that bitterness came some cruel statements — statements that later he probably regretted but was too vain to withdraw.

England had a strange effect on him — it both liberated and enslaved him.

There he grew — as one must in a land where the very air speaks of fulfilment. But in exile he remained chained to a vision of Ireland that became increasingly unreal.

And those who argue that the division of his plays into 'Dublin' and 'English' is too crude have a thin case.

The Dublin plays are less ambitious technically — but also give the impression of having been written with greater care.

'Cock-A-Doodle-Dandy' and 'The Purple Dust' have the old O'Casey exuberance — but seem as if they were written in a hurry and never corrected.

There are passages in both that are embarrassing — you wonder if the playwright has lost all sense of self-criticism.

'The Silver Tassie' is generally deemed the best of the later plays — and yet you feel that it could have been a great deal better.

Its second act is a marvellous expression of war's reality — but the acts set in Dublin are unconvincing.

Perhaps it is the fault of the language: it is hard not to suspect that O'Casey was not too sure about what he wished to achieve.

Some of the language is marvellous — it is, of course, unrealistic but only in the sense that it is exaggerated.

But there is much too that is thin and unreal. perhaps O'Casey meant it that way. But the effect is unhappy.

One leaves the play with the feeling that he did not achieve his purpose — and one remembers it for some of the parts rather than the whole.

But as a kink of dramatic essay to show the obscenity of war it works its passage.

'The Silver Tassie' has a message — but it conveys it in an unconvincing way.

War, of course, is generally obscene — and the 1914 war was particularly so, but we would apprehend this truth more sharply if the people in the play were more than symbols.

That, more or less, is all they are. The 'message' is infinitely better expressed in that modern classic 'And The Band Played Waltzing Matilda'.

There is far more reality in 'Red Roses For Me'. But it is a diffuse work — it lacks the clean outline of the 'great Dublin three'.

Yet it cannot fail to excite: its theme is both charming and powerful.

The image of the young man hungry for learning evokes a deep response from all of us.

We are excited by the yeast of youthful ambition.

And not only is Ayamonn flexing his spiritual muscles: it is the time of the Citizen Army and the country — or at least Dublin — is in ferment.

Yet it is hard to accept it as a great play: much of the language — even allowing for the nature of the times — is too forced.

It is only to be expected that a great playwright should expand and experiment — but there seems a vast carelessness about the later O'Casey.

'The Shadow of a Gunman' may seem slight and unambitious — and yet perhaps it contains a clearer and more relevant message than all his other plays.

And though its people may be symbols, they are also flesh and blood.

I prefer to think that it contains more of the 'real' Sean O'Casey than the later plays and the autobiographical volumes.

Few men are fitted to write their own life-stories: they shy away from the humiliation.

And God save us all from men who are deemed great in their lifetime — they almost always turn into bores.

And so at the risk of appearing alarmingly unoriginal I applaud the early Sean O'Casey — and have reservations about the sage of Devon.

Is there such a thing as a quiet country life?

A recent tragedy predictably evoked a flood of cliches about 'a quiet countryside' and 'a sleepy market town'.* They revealed minds that were not only stale but patronising.

A quiet countryside exists only in the mind of a town-dweller: life in the woods and the fields and the lakes and the streams is as fiercely competitive as in the urban jungles.

A tree has been well described as a tiger with only its tail above the ground; in the fight for living room it follows a hard imperative.

The thrush is among the gentlest of birds but it must kill to survive.

The trout is the loveliest of fish but it doesn't spare the fly.

Life in 'the quiet countryside' is an unending battle.

You are unlikely to hear Paris described as a sleepy city but its denizens probably spend more time in bed than do the people of Mullingar or Macroom or Magherafelt.

If a town looks deserted, it is probably because its inhabitants are away at work or in their gardens or in their kitchens passionately making jam.

Few cities appear sleepier than London on a Saturday night, mainly because few people live in its centre and because many of those who live in the outer areas go away for the weekend.

The urge to get away for a few days from the city is a comparatively modern manifestation.

It began with the coming of the railway and was accelerated by the proliferation of the bicycle and the car.

There was a factor other than the availability of transport: cities began to grow too big for their own good; people looked for little refuges from the concrete and the asphalt and the tarmac.

And in late nineteenth-century Britain this yearning for wood and field and lake and stream engendered a mild literary revolution — 'the country book' was born.

Publishers succeed by keeping both ears to the ground, if not simultaneously — and so fashionable writers were dispatched with nice commissions to wax lyrical about the wonders of rural life.

Most wouldn't know a rook from a crow or an alder from an elder or an eel from a lamprey.

This blissful ignorance was probably an advantage: the intrepid explorers saw what they wished to see; they weren't disturbed by the complexities of real life.

Edward Thomas christened this new breed "The Norfolk-jacket school of writing".

They tramped the rural lanes and the forest tracks and the bridle paths (aah, those bridle paths — they were the quintessence of country life) and felt themselves in unison with nature.

And, of course, they never strayed too far from some congenial inn where they puzzled mine host by insisting that supper should consist of ale and bread and cheese.

Most of the writing which came out of that mild wave is akin to what geologists call fool's gold — but there were at least three prospectors, Richard Jefferies and W. H. Hudson and Edward Thomas himself, who brought back the real thing.

The unity of their spirits was neatly symbolised by Thomas when he dedicated his biography of Jefferies to Hudson.

Technically Thomas was the best writer; Jefferies was probably the most knowledgeable; Hudson was certainly the most forceful.

He was born in Argentina of North American parents; he managed to avoid prolonged academic education; the Pampas was his school.

The time came when he had to seek gainful employment: he caught rare birds and sent their skins to an institute in Washington; eventually he extended his trade to London; his notes on the birds were published by The Zoological Society — thus he began as a writer.

In 1874 he set sail for Southampton — at the age of thirty-three he was about to invade the great world outside.

His early years were not noticeably successful; Jacob Epstein's famous sculpture, Rima, is said to mark the spot where Hudson used to sleep in Hyde Park.

He survived by resorting to the most desperate expedient known to civilised man — he married his landlady.

Seemingly the union was not a disaster; when Emily asked him why he referred to her as his companion in one of his essays, he said — "Not every wife is a companion".

From his base in Westbourne Park he set out to explore a new world; he chose The Wiltshire Downs as his testing ground — in its sense of openness and freedom no other part of England comes as close to the Pampas.

To make a living he wrote two novels; the second, 'Green Mansions', brought fame — and enough fortune to enable him pursue his vocation.

Twenty-five years of wandering and note-taking culminated in 1910: 'A Shepherd's Life' is a classic, if by the term you mean a book that deals so well with its theme that you cannot imagine better.

'A Shepherd's Life' is being republished this week. The title is a flag of convenience: Hudson combined wide and deep observation of life in Wiltshire and beyond with the memories of a man who had spent a long career tending sheep on the downs.

His name was James Lawes and he lived in the village of Martin; the author called him Caleb Bawcombe and the village Winterbourne Bishop — it was a harmless concession to public taste.

The life of a shepherd in those days was an epic of loneliness; he might go for weeks and months without meeting another human being between dawn and dusk.

So rare were chance meetings in the life of James Lawes that he used them as time-marks: "It was two weeks after the day the stranger sat down and smoked a pipe with me on the plain . . . "

A shepherd needed to know all that was going on about him; almost inevitably he became wise in the ways of those beings that shared the downs and the plains with his sheep.

And, most of all, of course, he needed to understand his dog.

Thomas Hardy believed that every sheepdog lusts to kill his charges; he could have been right.

What is certain is that much of a sheepdog's instincts are restrained almost to the point of extinction: he mush not, for example, waste time in chasing hares and rabbits — thus he acquires 'an artificial conscience'.

The tales of Shepherd Lawes weren't all about the creatures of the downs and the plains.

There is the story of the gamekeeper whose wife longed for a child — in vain.

One morning he found great difficulty in opening his front door; a basket, complete with new-born baby, was suspended from the latch; inside was a note saying:

Take me in and treat me well,
For in this house my father dwell.

Despite the eccentric grammar the gamekeeper's wife was overjoyed. She asked no questions. The child was christened Moses Found. The story sheds a little light on rural ethics — and on the origin of surnames.

Not all the stories that Hudson gleaned in the rolling countryside under White Horse Hill are as indicative of the good in human nature.

Shepherd Lawes and his coevals remembered the terrible years when the farm labourers were worse off than slaves — because slaves at least had security.

163

You will find little reference to 1831 in the conventional books on English history; the uprising of the farm labourers is deemed as belonging to sub-history.

It was a sad little revolt, inchoate and hopeless; its leaders, such as they were, had enough sense to know that it couldn't succeed; essentially it was a cry of rage from a cruelly-oppressed people.

The last straw — and certainly there is no pun intended — was the coming of the threshing machine; for thousands who wielded the flail it meant that they would be out of work in the dark months of Winter.

The welfare state was then hardly even a dream; the unemployed labourer and his family faced starvation.

The most remarkable aspect of this uprising was that the insurgents confined their violence to wrecking threshing machines and burning ricks — not one of their oppressors was harmed.

That admirable show of discipline did little good for those who were rounded up: in Salisbury, one of many towns where Special Commissions 'tried' the insurgents; thirty-four were sentenced to the gallows — and many more sent to Australia.

"When the trial was ended and those who had been sentenced were brought out of the courthouse to be taken back to prison, from all over The Plain and from all parts of Wiltshire their womenfolk had come to learn their fate. . .

"They were gathered, a pale, anxious, weeping crowd, outside the gates. The sentenced men came out looking eagerly . . . until they recognised their own and cried out to them to be of good cheer . . ."

It was hard for your folk to be of good cheer when you were due for the gallows.

At least one of the condemned men seems to have had a sense of humour, black though it may have been: "Don't cry, old girl . . . 'Tis only fourteen years I've got — and maybe I'll see you all again."

Very few came back from Australia; how their families managed, only heaven knows.

The authorities got over the difficulty of producing evidence by a simple ploy: "The smell of blood-money brought out scoundrels who for a few pounds were ready to swear away the life of any man."

One story especially illustrates the climate of the time.

In the course of the uprising a farmer shot dead a labourer — there was no doubt about the murderer's identity but he was never brought to trial.

A witness told the inquest that the subject of their enquiries had been shot by a yeoman whom he had attacked with a stick; it was a rather unlikely story, especially as the yeomanry didn't arrive on the scene until after the man was dead.

164

Even in death the poor weren't immune from injustice.

* Hungerford, the scene of last week's tragedy, is in the very heart of the countryside that Jefferies and Hudson and Thomas loved so well.

Evening Press, **Tuesday, November 7th, 1978**

Dylan Thomas — Water Poet

What is a nation? Perhaps the best answer is this: the word once had a fairly exact meaning; it no longer has — but because it survives, we tend to use it even though aware of its unsatisfactory nature.

But why does it survive? Perhaps the best answer is this: even though it has more or less lost its descriptive use, it is still a word we can hardly do without; it indicates, however imperfectly, a spiritual entity that is none the less real because exceedingly difficult to define.

There was a time when a community of people was a nation—when they shared a common home and a common way of life and when their wealth was not largely the property of one class but for the common good.

It is doubtful if any such society exists in the world today, at least on a scale big enough to warrant the name 'nation.' Yet the word persists.

And now and then on occasions of great collective emotion you feel that it not only persists but glows with indestructible life.

At Croke Park, for instance, on the day of an All-Ireland final, the emotional flood released by the National Anthem engulfs you—and at that moment he who would question the existence of an Irish nation would seem a monstrous heretic.

But cool analysis would at least give him the right to state his case — that the occasion could hardly be called national when a significant section of the supposed nation was only symbolically represented.

And at a rugby international in Lansdowne Road, the anthem arouses a powerful sense of nationhood too—and on this occasion there is no section unrepresented.

But again cool analysis whispers that the common bond is only for a little while and in a cause that is of only illusory importance.

How then do you define a nation? Will you say that you cannot define it but can feel it in your blood? The blood in that sense is not to be ignored — but divorced from the intelligence it is a ship without a rudder.

Twenty-five years ago this week Dylan Thomas went into the good night. Since then he has grown in folklore as the archetypal poet — and as the archetypal Welshman.

And yet how Welsh was he? How Welsh could he have been? The image of Wales as a tight little fortress wherein dwell the old dark race driven thither by the Saxons is hardly justified—it is a land made up of five fairly distinct parts.

Dylan Thomas was born in one — the industrial strip between the mountains and the Bristol Channel; he had close ties with another — the farming-and-fishing country of the south-west. Yet he hardly 'belonged' to either.

Or at least neither seems to have given great sustenance to his poetry.

In his day the industrial heartland of Wales was a world dominated by enormous energy: to pass through it in the train in the darkness of morning was an unforgettable experience — the glow from the steel mills and the blast furnaces and the hurrying of car lights to and fro gave a tremendous sense of power and purpose.

And it borders another Wales that might belong to a different age: the south-west is an unvisited country, even though the Paddington-Fishguard train goes through it.

The old language lives there. Life in its small farms is hard. Even today you can see women out before dawn attending the tide to gather seaweed.

You will find little of the throb of the industrial south-east in Dylan Thomas — in Daniel Corkery's famous test he would hardly have been an integral part of the crowd at an international game in Cardiff Arms Park.

Nor is there much of the south-west: you can pass through that land and, apart from one celebrated phrase, it will not bring his poetry into your mind.

And yet there is something in him that seems quintessentially Welsh. An enemy would call it rhetoric — a friend would call it passion.

R.S. Thomas in one of his poems says that there is no Wales, only myths and ruins and dreams. There is no more intelligent poet — and that concept, wayward and self-pitying though it may seem, is not to be dismissed lightly.

And yet his own work contradicts it. A Wales lives in his poetry — it is the Wales of the inland hill country where bareness and sheep and the old language are the elements that impinge on the passer-by.

Thomas expresses it most powerfully in his poem about the death of one of his parishioners.

You remember Davies? He died you know,
With his face to the wall, as the manner is
Of the poor peasant in his stone croft
On the Welsh hills. I recall the room
Under the slates, and the smirched snow,
Of the wide bed in which he lay,
Lonely as a ewe that is sick in lamb
In the hard weather of mid-March . . .

When you read that and then read say, "Fern Hill' or 'Poem on his Birthday'; you can hardly believe that the two Thomases were born into the same time and the same small country.

But what R.S. Thomas means when he says there is no Wales is probably that there are several—and that the bond that makes them one is impossible to divine.

And yet perhaps it is not . There is another well-known poem of his that seems a finger-post. It begins:

When I was a child and the soft flesh was forming
Quietly as snow on the bare boughs of bone,
My father brought me trout from the green river
From whose chill lips the water song had flown.

The other Thomas might have written those lines.

The bond then is one of language? In a way it seems to be, even though the Thomas of the hill country is greatly influenced by the old tongue — and the other Thomas knew it not.

And it is not untrue to say that there lives in Wales a culture that tends to unify and in a sense make democratic the whole country.

A passion for music is part of it. The piano in the miner's front room and the beautiful singing of the Welsh crowd before a rugby international are not mere surface show — they are the flowering of a love that is very deep.

And there is a great respect for language — the motto of the school where David Thomas, Dylan's father, taught in Swansea was 'Virtue and Good Literature'.

And at least in Welsh-speaking Wales there is a long tradition that looks on poetry as something as much a part of life as the postman and the weather and the chapel.

Much of what passes for poetry there might be better described as the product of brilliance in prosody — but at least it indicates a delight in weaving verbal patterns.

And it is not too fanciful to think of Dylan Thomas as by no means untouched by the love of music and of poetry that were part of his heritage, son of Swansea though he was.

And there was a nice irony in his leap to fame: he was embraced because his poetry seemed so fresh and spontaneous to a generation tired of W.H. Auden and his cackling brood — fresh it was but so cunningly sculpted that few suspected the presence of an armature.

There was nothing wrong with that: it was legitimate cunning — but Thomas's ability to make words do his will was also his greatest flaw.

He developed a way of speaking: it was wonderfully apt when the emotions and perceptions and concepts it expressed were fresh and strong — in time it became a vividly-painted wagon carrying a trifling load.

He had no deep roots: the only reality in his life was his childhood. He never grew out of it — for the rest of his days he was an exile in the kingdom of words.

And when that mine of childhood had been worked out, there was little left for him to sing.

No one knew that better than himself: there is no need to regret his dying young — his death was far more a suicide than that of him who shuffles off the mortal coil in one fell swoop.

Dylan was the great bard of youth. One thinks of Coleridge's definition of genius — the child's vision in coalition with the man's power. But the power in Dylan was mainly expertise with words.

And when the well of youth went dry, the verbal wizardry became intensified as a compensation. The marvellous word-weddings gave way to made marriages. Love faded into calculation.

The early Dylan was a true poet, a maker, one who made a cunning net of words and swaggeringly hauled out the fish too elusive for others to capture, who divined the hidden waters and sculpted them into brilliant fountains.

The later Dylan was an unwilling charlatan, a deviser of bright puzzles for the insatiable maw of the American thesis factories. There is even, God between us and all harm, a big book that gives a page-facing-page 'translation' of the later 'poems'.

The truth, as Samuel Johnson was wont to say, needs not so much to be discovered as repeated — and it is with the air of a man announcing that Rugby is among the major passions of South Wales that one ventures to say that Dylan Thomas's genius is distilled into a hatful of poems.

The language in those is as bright as a trout, as flavoured as country butter, as clean as an ash tree, and it contains elements of the strong man and the acrobat and the juggler.

In the mustard seed sun,
By full tilt river and switchback sea
Where the cormorants scud,
In his house on stilts high among beaks
And palaver of birds
This sandgrain day in the bent bay's grave
He celebrates and spurns
His driftwood thirty-fifth wind turned age;
Herons spire and spear.

And 'Fern Hill' perhaps is his most satisfying poem. It was surely a great oasis in his loneliness.

'Time held me green and dying
Though I sang in my chains like the sea.

And there is, of course, 'Under Milk Wood', and if Dylan Thomas ever came near finding a lasting oasis it was in that village in West Wales that he metamorphosed into literary folklore.

It is a water village, webbed by river and sea. And Thomas, like every child, loved water.

Land is stern and seems timeless — it has been there before you and will be there after you. Water seems the embodiment of imagination, malleable, ever-changing — always being re-born in another place, another shape.

'Under Milk Wood' is lovely and will go on delighting generations — but it is sentimental.

It is a piece of confectionery and has its place. R.S. Thomas would have left a very different picture — though he too might have spoken of the 'heron-priested shore'.

Dylan Thomas's tragedy — if you can call tragic his short voyage into immortality—was that all his life he lived by words. He never knew the corrective of a job where the worth of your work is objectively tested.

He was a master of words who eventually became their slave.

But the bright stippled fish remain swimming in that keep-net — and when you read the good Dylan Thomas, you are 'green and golden', happy 'in the sun that is young once only' rejoicing in 'so few and such morning songs'.

Evening Press, **Tuesday, January 30th, 1979**

Robert Burns — O' Mice And Men

I thought of Chatterton, the marvellous Boy,
The sleepless Soul that perished in his pride;
Of Him who walked in glory and in joy
Following his plough, along the mountain-side.

Thus wrote William Wordsworth in 'Resolution and Independence.'

Had Robert Burns, the second subject in that little eulogy, lived to read it, he would have come up with a comment chiselled in ribaldry.

Part of Wordsworth's attitude towards nature came from his marvellous ignorance of it — it might be better to call it a fruitful innocence.

Burns grew up at a time when farming was depressed, and especially in Scotland—and on a farm that in the best of times would be a sour wife rather than a smiling mistress.

His father entertained no illusions about the good earth — he was only a tenant anyway and did not wish to see his sons enslaved by both landlord and poor soil.

He saw to it that they got the best-available education. And that was at the feet of a young man called John Murdoch — he held classes from house to house — he was a kind of Scots hedgemaster.

The concept of hedgemaster has been romanticised too — but it may be true that their schools were the best, if only because their pupils tended to be a few years older than their counterparts in regular academies.

And the belief that the Scots have a hunger for learning is no myth. The image of the student returning to the university with his sack of oatmeal to last him for the term is well founded.

And their excellence in the professions, especially in medicine and engineering, is rooted in hard work. So is the poetry of Robert Burns.

And yet how different is the popular image. In folklore he is the untutored genius who warbled his native woodnotes as freely as a thrush and who came to an early grave through his bibulous and amorous propensities.

The truth is that he drank but little and had a poor head for the stuff — and he was a timid lover whose number of affairs was probably less than the national average.

From youth he was afflicted by the after-effects of rheumatic fever, a legacy of the cold and the wet of those fields through which he was supposed to have walked in glory and in joy.

But it is doubtful if it was the direct cause of his death. He died at the same age as Dylan Thomas and in curiously similar circumstances.

It is now fairly well established that Thomas did not die from the effects of an epic drinking bout but was shot down by an injection of a substance that was new and fashionable in 1953.

Burns was told by a bright young doctor that there was an infallible cure for his pains — bathing in the Solway.

He dutifully followed instructions — and after a fortnight never suffered the slightest pain again, at least on this terrestrial globe.

But folklore will not tolerate such passive exits — and Thomas and Burns went out roaring from drink.

The reason for the myth is simple: people will accept genius only when it is accompanied by some Hamlet-like flaw.

There is another side to Burns in folklore too: he is seen as a bold rebel against despotism, a gallant apostle of the rights of man.

The irony is that it was true of his better self — the Robert Burns he would like to have been.

But it most certainly was not true of the day-to-day Robert Burns: when his sympathies with the French Revolution put his job as an excise man in danger, he wrote to the authorities what must be the most craven letter ever penned.

Nor was he a gallant lover. You will find that lack in his letters too: the poet of 'Ae Fond Kiss' and 'O Wert Thou in the Cold Blast' could give way to the social being who might charitably be described as a male chauvinist jackal.

Nor could anybody with the least sense of logic claim that he is the national poet — at best he was the voice of the Scots lowlands.

Who then was Robert Burns? What substance is there when all the accretion of folk wish and dream and sentiment is cleared away?

One is tempted to invoke a familiar image — that of a tree being like a tiger with only its tail above the ground.

There are poets who seem to owe nothing to the place and time in which they grow up, who seem to have no heartland.

And 'surroundings' in this case signifies more that the present: it includes the power of the past to press in on the immediate like the dark around a lighted house.

It may seem strange that so poor a place and so depressed an era could nurture genius — Burns grew up at a time when his father and his neighbours were not unlike the poor whites of the American depression.

One of the fruits of empire is cheap food from the colonies — and for the less fortunate in the mother country that means the choice of living in poverty at home or dying for the flag in a foreign clime.

And Scotland came to know the evils of distant government: the bureaucrats of the day had it marked down to be a land of grouse moors and fishing lochs and sheepwalks.

And yet Robert Burns was not unlucky in his place and time: like his Irish contemporary, Eoghan Rua O'Sullivan, he inherited a world where the spirit had not abdicated.

It may not be true that the peasants of 18th century West Munster were fluent in Latin and Greek; some of them were — the point is that most would like to have been.

There was a similar love of learning among the Lowlands Scots — and like their West Munster counterparts their mental barns were well stored with poetry.

You will remember Tam O'Shanter. As he rides home that night of wind and rain and thunder and lightning, Burns has him.

Holding fast his gude blue bonnet,
While crooning o'er some auld Scots sonnet.

And yet this heritage put Burns into a kind of trap: he was educated — but could not afford to go on to a profession.

And his life became a thing of false starts — he was like a jackdaw that begins building in a chimney and sees his work pounded down into the fireplace.

He failed in two farms; he thought of emigrating to Jamaica; his first two affairs of the heart ended in extreme public humiliation.

And yet there was something in him that poverty and poor health and economic and social failure could not suppress: the river of his spirit might be diminished to a trickle — and yet could suddenly overflow.

In short, he was a poet: he carried with him the philosopher's stone that could transform seemingly base metal into gold.

The poet is heir to the primal word-maker: he may not be the most successful hunter — but when he captures an animal, he captures it eternally.

And Burns was also a genius: as surely as Columbus did, he discovered a world that hardly anybody else believed to exist.

The poetry he inherited seemed to speak of a romantic age — it hardly seemed possible that poetry could be born out of the Ayrshire he knew.

Even today when farming is relatively prosperous and Prestwick Airport has made the wide world seem nearer, it is a lonely country. In Burns's time it must have seemed a prison to a young man of spirit.

By staying in it he escaped from it. And his key was his astonishing confidence in the language of the fields in which he grew up.

That dialect was then despised by the formally educated and not highly thought of even by those who spoke it.

And it is certain that Burns did not make the decision without a struggle — but once he had made it, he did not hug the shore.

And again he was lucky: a century later the Dorset poet, William Barnes, decided to write in his local language — and can hardly be said to have succeeded.

It may be because he was the less a poet — a possibly greater reason was that the Dorset dialect was not a good medium for poetry. It was Thomas Hardy's childhood tongue — but he abandoned it.

Or perhaps it was that Lowland Scots was more generally comprehensible. And yet one feels that it is a remarkable language.

Part of the attraction of the Border Ballads lies in their power to express physical reality — Lowland Scots seems to have that quality to an even greater degree.

When you read 'To a Mouse' for the first time, you may not understand all the words in it — but you apprehend their meaning.

And the impact it makes would be almost impossible in conventional English.

And there is another great element in that language too — its precision.

That may surprise some — they look at dialects as intellectually inferior.

But even this present day in Kerry, after 16 years of television's blandness of language, that same instinct for precision survives among the country people.

You may come home from the bog in the evening, for example, and be told by someone: "I discerned you coming over the brow of the mountain."

And a man narrating something will sometimes stop and wait to let the best word come — and his company will know exactly what he is doing.

Robert Burns, of course, was more than a passive instrument — he was a maker of language too. And it is salutary to remember that 'poet' originally meant 'maker'.

Not all the words he used were common coin in his day: he brought them back into the mainstream of the dialect and, in a sense, immortalised them.

Some of them are Latin words used in their original sense—an indication that all mankind is one.

And, of course, Burns was very aware of the conventional English poetry of his day—and was by no means unaffected by it.

"Tam O'Shanter" is a rollicking piece of Scots ribaldry—but it was meticulously worked out. Alexander Pope would have approved. The precision of its language is consummate.

Robert Burns fought a brilliant campaign against the negative side of puritanism — its positive side is exemplified in his work: his output in his glorious year, 1786, would seem sufficient for a long lifetime.

His work as a collector and refurbisher was tremendous—on it, of course, his popular reputation largely rests.

And, no doubt, over the drams and the haggis last week the 220th anniversary of his birth was washed around with the familiar songs.

Did anyone recite "To a Mouse"? If Burns had never written another line, that poem would have made him immortal.

That wee bit heaps o' leaves and stibble
Has cost thee many a weary nibble.
Now thou's turn'd out, for a' thy trouble,
But house or hauld,

To thole the winter's sleety dribble,
An' cranreuch cauld.

You will hear it said that Robert Burns would have been a greater poet if he had lived to maturity.

That belief is not well founded: his best work was done by the time he was 27. Young men in poor health mature rapidly: they fear that their lease has a short date.

That marvellous harvest of his 28th year brought him great fame and some fortune.

It is doubtful if he found much comfort in his new job as a gauger—the excise man was away with the de'il.

Robert Burns's life was a tangled struggle. But "he sang in his chains like the sea." In a sense Wordsworth was right.

Evening Press, Tuesday, June 14th, 1983

The brave sad voyage of Edward Thomas

James Ashcroft Noble was a member of a species that flourished in the nineteenth century but is now almost extinct—he was 'a man of letters'.

Even the label is almost forgotten — in Noble's day it was a badge of honour.

It denoted someone who had a passion for literature and wrote about it with loving care.

That care included an unfailing respect for grammar and syntax—and a reluctance to use a word when a better one was signalling, however faintly, from the mind's forest.

Richard Church was probably the last of the breed.

And the demise of 'John O'London's Weekly' was like a declaration that an era had ended.

That great little magazine set high standards — and maintained them.

It died from a familiar disease — lack of advertising revenue.

The ad-men knew that most of its readers were elderly — and unlikely to have numbered accounts in Swiss banks.

And it could not survive on advertisements devoted to secondhand books and home-made jam and boarding houses in North Wales.

James Ashcroft Noble dwelt in south-west London, then predominantly rural.

Even today it hasn't lost that character.

It is in the main a world of modest houses; almost all have little assiduously-tended gardens — and almost every garden has a greenhouse.

And the profusion of trees accentuates the rural flavour.

Helen was the middle of James's three daughters—and had the romantic outlook often found in second children.

The origin of this is probably easy to explain: the oldest child is often cast in the role of little father or mother and hence likely to imbibe conventional values; the second child is less manacled by the chains of common sense.

Helen Noble was lucky in her heartland.

It was near enough to the metropolis to be affected by the great world that William Wordsworth attempted to capture in his sonnet 'On Westminster Bridge'.

And it was rural enough to be a wonderland for a young person who loved birds and flowers and trees.

It is likely that James Ashcroft Noble would now be forgotten but for his association with Edward Thomas.

The man of letters loved to encourage young talent, probably because his own early career had been unaided: he divined the promise in Thomas's fledgling efforts — and befriended him.

And thus the apprentice writer met the young girl of romantic mind: they came together as naturally as two streams in the same catchment.

Noble, not surprisingly, encouraged the friendship: his wife disapproved; no doubt she thought it was enough to have one man of letters in the family.

Her disapproval, of course, strengthened the bond — and when she forbade the lovers to meet, she sealed it.

Edward and Helen were married in the last year of the century: she was twenty-three—and he was twenty-one.

We tend to believe that in this generation we are experiencing a sexual revolution — Helen Noble's story suggests otherwise.

She disapproved of marriage — and welcomed the prospect of becoming an unmarried mother.

Thomas agreed in principle — but felt that he had already caused his family enough pain.

It was his only brush with compromise: in his ambition to be a writer he was unbending.

He was still at Oxford—he left the following summer with a second-class degree.

It was a bitter disappointment to him—even though he knew he was in distinguished company. Samuel Johnson, among others later to become famous, had left the university, without a degree at all.

Thomas, however, had gone up on a scholarship and had been determined to make his mark as an academic: possibly he became too involved in the general life of the college; he left behind a reputation as a good oarsman and a first-class drinker.

Now he had a wife and a son—and no job. The Civil Service beckoned—but he was determined to be a river rather than a canal.

At least that was how he saw it—ironically, some of the greatest writers were civil servants all their working lives.

His madness was not without method: his essays had been welcome in so many magazines that he saw a golden future—he was wrong.

'Plain living and high thinking' was an attractive slogan—but the plainest of living demands money. And when Thomas applied to Henry Nevinson of the Daily Chronicle for work, the famous editor saw that he was grievously undernourished.

It was a strange interview: when asked what work he could do, he said 'none'. And when asked what subjects most interested him, he boasted that he knew nothing about anything.

Nevinson was nevertheless impressed—and gave him a start as a reviewer.

Other papers gave him similar work—but his troubles were only eased. Reviewing was badly remunerated—mainly because it was the province of those for whom their name and words in print were a reward more important than money.

Between August 1900 and June of the next year his earnings totalled £52.

Much of his time was now spent walking—or rather trudging—from the office of one paper or magazine to another.

And often he came home as weary as a fox that has been out all day and caught nothing.

The romantic dream of being a writer proud and free was like a delicate flower that had come above the ground too early and been singed by the frost.

And to speak of him coming home is an exaggeration: Edward and Helen and their son lived in a few cheap rooms in a part of industrial London old enough to be a slum.

The idyllic days of their courtship when they roamed the countryside seemed part of an irretrievable past.

Thomas had a greater worry than his failure to make a decent living: he was beginning to doubt his talent.

In his student days he had spoken of his essays as his 'golden apples' — as his judgement matured, he suspected that they were gilded rather than golden.

The suspicions weren't unfounded: he was too much under the influence of Richard Jefferies and Walter Pater.

Jefferies is rightly regarded as the greatest of English writers about nature—but his most influential book, 'The Story of My Heart', is marred by synthetic mysticism.

However, much of his influence was for the good: he taught Thomas, among other things, the importance of skinning his eyes and seeing the world for himself.

Pater's influence wasn't altogether detrimental; he believed that prose is not poetry's poor relation—but in his zeal he tried to make language into something like abstract art.

Today his writing has the faded beauty of a flower long pressed between the leaves of a book.

And yet despite the palpable influences, those early essays, the golden apples, can still be read with pleasure—behind the occasional affectation there is a strong and pure mind.

Thomas carried self-criticism too far.

The finished work never seemed to match the excitement that had engendered it.

And the words he had uttered in Henry Nevinson's office were more than a crazy form of humility.

Thomas understood what Coleridge meant when he said that he felt himself to be an involuntary charlatan.

And as early as his mid-twenties he would have echoed Daumier's verdict: "The wedding feast of the world has passed me by."

Now he had a daughter—and two small children put a strain on his meagre earnings: the welfare state was still only a Fabian dream.

The gentle dreamer whom Helen Noble had married became prey to prolonged depressions during which he savaged those dearest to him.

Helen, of course, suffered most of all. Her sweet and strong nature seemed unaffected — but the bitter words though forgiven were not forgotten.

And it hardly eased her mind to discover that he kept a revolver in secret—whence it came nobody ever found out.

In this he resembled Fred Archer, the great jockey, and Randolph Turpin, the great pugilist—in their case the secret was discovered too late.

A move to a cottage in Kent eased his suffering—and for the rest of his life as a writer he was to live in rented rural houses within striking distance of London.

Thomas's heartland was the south-east; there is no more maturely lovely part of England. But he did not look on it sentimentally—he knew what had made it so.

His love embraced the people who had hewed and dug and ditched and drained and herded and milked and hedged and paved and planted and reaped and done all the other land-tasks for generations.

And when he volunteered for war, he knew what he was fighting for — it was something far more tangible than a flag or an anthem.

That smooth deceiver called hindsight tends to blind us now to the realities of 1914 — we forget the enormous prestige enjoyed by the German army then.

It had distinguished itself in the Napoleonic wars—and then smote France like a thunderbolt in 1870.

The idea of Britain being occupied was far from fantasy. And Thomas fought to prevent that.

And the pacifist stance was dishonest—it could not be maintained if sufficient numbers had not taken the other road.

Sometime in 1915 Thomas attempted to express his feelings about the war in a poem that in some ways is most untypical of him.

Of all his published poems it comes nearest to failure—it stands up because of its passion.

> . . . *Beside my hate for one fat patriot*
> *My hatred of the Kaiser is love true.*
> . . . *But with the best and meanest Englishmen*
> *I am one in crying, God save England.*
> *She is all we know and live by, and we trust*
> *She is good and must endure* . . .

That may smack of rhetoric — but it is plain enough. Thomas did not own a house or a field—but yet he feared that he might be dispossessed.

He was like a fish sensing the coming of a pollutant to his native stream.

Edward Thomas was very acutely aware of man's need for spiritual territory—for a mental climate and for streets and fields where he knows himself to belong.

And in England that climate has tangible forms—no country is more blessed with open spaces that are the people's own.

Its commons and downs and heaths and moors and forests give "freedom" a precious dimension—and that dimension is fiercely guarded by those to whom the sweetest words in the language are "right of way".

Thomas was the poet of that world, the pilgrim who delighted in walking the paths that generations had beaten out.

The mystic best knows reality—and Thomas felt that in travelling along those ancient ways he came nearer to the soul of his ancestors.

The honesty that sent him to war is present in everything he ever wrote: no one stands up better to Ernest Hemingway's test — that nothing should go bad afterwards.

And there was not an iota of sentimentalism in his courting of the common people—he knew that they, unlike the various intellectual elites, could not afford to be triflers.

For countless generations they had worked the land—and without an amplitude of strength and toughness and wisdom they could not have succeeded.

And he believed too that they were the true begetters of language — others might polish it and package it but they were the miners who hacked out the words at the coalface.

He felt that the precision of their language came from the demanding nature of their life—and that its colour was the kind of poor man's poetry that was their anodyne.

One of Edward Thomas's great ambitions was to write a poem that would encompass all he knew and sensed about the people of the fields.

He didn't live to achieve it—but "Lob", a poem of 150 lines, is a kind of sighting shot.

Lob, its hero, is the archetypal country labourer. Thomas imagines him as an immortal, who

> *Although he was seen dying at Waterloo,*
> *Hastings, Agincourt, and Sedgemour, too*
> *Lives yet . . .*
> *He first of all told someone else's wife*
> *For a farthing she'd skin a flint and spoil a knife.*
> *And when she heard him speak*
> *She had a face as long as a wet week.*

He once talked with Shakespeare in the hall, having borne in the logs when icicles hung by the wall.

It was he who christened the rose campion Bridget-in-her-bravery — and called cuckoo-flowers Milkmaids and wild clematis Traveller's joy.

> *And*
> *He never will admit he's dead*
> *Till millers cease to grind men's bones for bread.*

Edward Thomas did not think of the town as hostile to the country: he deemed Wordsworth's "Westminster Bridge" as much a nature poem as "The Immortality Ode".

He would not agree with William Cowper's "God made the country and man made the town" — man had made both. And he was not pleased to be known as a "country" writer.

For him the country was fascinating because there you could see man's intelligent co-operation with nature.

In his own unstrident way he was a socialist—one of his greatest poems, "The Owl", contains the essence of the socialist creed in a few lines.

In it he tells how he came to an inn after a winter day's walking and had food and fire and rest. Then out of the night came an owl's cry, "a most melancholy cry".

And salted was my food and my repose,
Salted and sobered, too, by the bird's voice
Speaking for all who lay under the stars,
Soldiers and poor, unable to rejoice.

Thomas had much in common with the "soldiers and poor". Even though he worked very hard as a writer of commissioned books, he never knew even the most modest security.

And it is doubtful if in his heart he desired it — he sensed that only hardship brings understanding: he would look for wisdom to a gypsy rather than a don.

And he admired the men who lived without counting the cost, the wild men who could put in an epic day's labour after an epic night's drinking.

The poem called "A private" is a little requiem for such a one.

This ploughman dead in battle slept out of doors
Many a frosty night and merrily
Answered staid drinkers, good bedmen and all bores:
"At Mrs. Greenland's Hawthorn Bush" said he
"I slept" . . .
And where now at last he sleeps
More sound in France — that too he secret keeps.

Edward Thomas was not a staid drinker nor a good bedman — in his own quiet way he was a rebel against conventional life.

He paid the price in life-long penury and in the bouts of depression that racked him—brought on by feeling that he had failed in the eyes of the world and in his own.

Once when asked by friends to consult a psychiatrist who was then all the fashion in London, he summed himself up: "He might cure me—but then I wouldn't be Edward Thomas anymore."

The south-east of England is to education as the country around Newmarket is to racing: it is as dotted with schools as Brighton is with pubs.

Not all of them are just ordinary schools dedicated to learning and to games and cherishing vague ideas about producing good citizens.

That part of England has long had a mysterious attraction for pioneers who wish to lead mankind onto a higher moral — and intellectual—plateau.

In the early years of this century Bedales School was the Camelot of progressive education—its very name was like a bell that evoked an image of man's nobler nature.

It need hardly be said that when the war came in 1914, not one of its staff volunteered for service—to a man and woman they were pacifists.

Such high-mindedness was admirable. But even better was to come: they all, to a man and woman, dedicated themselves to helping those whose menfolk were in the army.

And so one night they convoked a meeting in the village hall and assured the local women that they would dig their gardens and plant potatoes and cabbage and carrots and parsnips — and possibly leeks and cress and mustard as well.

It was a noble gesture—but, alas, a gesture it remained. And so the wives and mothers and sisters of the men who had gone to war got out their forks and spades and dug for victory.

Edward Thomas was one of the local men who had gone to war—or at least he was local in the sense that he lived in a house owned by Bedales.

In return his wife helped out in the school — and, no doubt, acquired great spiritual benefit in that noble environment.

When her husband was killed in France, she went into a long mental decline and was unable to work—and she and her children were evicted from their house.

Neither the story of the gardens nor the eviction would have surprised Edward Thomas: he had long learned to be underwhelmed by "noble" aspirations.

His intuition of man's story was that the most important part of it had gone unrecorded — below the surface that we call history lies the quiet heroism of countless generations of obscure men and women.

No one ever loved the English countryside more than he—and true love is not blind: it is unremittingly perceptive.

Thomas, unlike many of his contemporary writers and artists, had no illusions about rural life.

He was acutely aware of the harsh life of the rural labourer—and yet would not agree with the unrelenting pessimism with which George Crabbe depicted the life of the fields and the coast.

181

He knew from intimate experience how the mill and the factory and the urban slum diminished people.

The farm labourer was a 'wage slave' too—but in surroundings that buoyed rather than oppressed his spirits.

Thomas was to become something of a wage slave himself.

As his name as a man of letters grew, he got a stream of commissioned work.

He edited volumes of poetry, wrote biographies—and even guide-books for tourists.

His output was amazing. It would have been remarkable for a writer who never changed a word—Thomas was over fastidious: he was, after all, a man of letters.

The work was meagerly rewarded: publishers have never been in danger of being mistaken for philanthropists.

It also ate up time: Thomas felt like a man travelling through a beautiful countryside—but unable to lift up his head and enjoy it.

And yet he still managed to produce the occasional golden apple—and kept his sanity, more or less.

The countryside was his psychiatrist: the old rich world of southern England knitted up his sleeve of care.

Whenever he could afford the time, he went on long solitary walks.

Helen suffered. Often she had only the company of her two small children.

She had longed for a big family—penury decreed otherwise: eight years had passed since the birth of their second child when their third and last, another girl, was born.

Thomas by then had a moderate fame: he was well known enough to be adopted by a set who were in love with the idea of art but not with the hard grind necessary for it.

Thus he met Eleanor Farjeon, a writer of books for children, whose only link with the set was that her brother was its leader.

She became his confidant — and Helen's too. And Thomas found himself loved by two women—and all the more burdened by guilt because he couldn't love either.

He couldn't — because he couldn't love himself.

He was over conscription age when the war erupted—but he enlisted.

There is no reason for disbelieving his expressed motive: he wished to safeguard the England he loved.

It is possible too that he was tired of words and longed to do something that could be objectively assayed.

He proved to be a model soldier. He was among the thousands of British that perished in the Battle of Arras in the Spring of 1917.

Van Gogh the fire within

In a strange way Vincent Van Gogh was like the little boy who refused to be one with the crowd that admired the emperor's magic garment and insisted on saying that he was dressed in nothing at all.

Van Gogh saw the poorest and the most wretched as being worthy of admiration as much as emperors; he perceived the magic garments that the rest of the world could not.

It is easy enough to understand why his vision was so democratic: his own life was so consistently riddled with failure that he saw himself as one of nobody's people.

'He used to go and sit on the slag-heaps' said one of his colleagues in his missionary days, 'making drawings of the women while they were picking up coal or as they went away loaded with sacks. We noticed that he never drew bright things, those that we take to be beautiful'.

These words seem strange now when every other bedsitter in London and elsewhere is illuminated with a print of 'Sunflowers' or perhaps 'Road with Cypresses' or 'Van Gogh's Chair'.

It is one of the many contradictions in the life of a man whose popular image, due in part to Hollywood, is that of the archetypal romantic painter — but whose days were spent in that kind of patient labour familiar to the coalminers and the peasants he depicted.

Or you might compare his life to a jackdaw's method of making a nest: he first finds what he deems a suitable chimney and then goes about furiously gathering sticks and dropping them down.

This may go on for a long time until eventually a stick so falls that it lodges; thenceforth the odds are in the jackdaw's favour — and fulfilment is at least possible.

Occasionally a stick seemed to have lodged for Van Gogh — what tended to happen then was that he increased his fury and dislodged it.

Einstein it was who said that he who does not know the meaning of the word 'failure' should consult a dictionary — and despite Bruce's spider and Poor Richard and The Readers' Digest he was right.

People do fail — and in a way that allows no room for hope or excuse. Vincent Van Gogh failed in not one but many quests. His tragedy was that he died thinking he had failed in all.

He was born in Holland in 1853 in a little town near the Belgian border; his father was a clergyman — something he shared with many a child destined to be famous, including Nelson and Jesse James.

His birthplace, Groot-Zundert, is set in melancholy moorland — that it affected one who had an extraordinary sensitivity to surroundings is probable. Vincent grew up a solitary child who alternated between high elation and despair.

A great many essays have been written to explain his 'madness' — most ludicrous are those that trace it to an injury at birth: the truth is that he was too sane, too perceptive, too uncompromising.

If he had a flaw, it was that he lacked humour, but humour is sometimes a mask for defeat: the messiahs, those who believe they are born to change the world, are generally untrammelled by it.

It did not unduly impede Van Gogh: he was surprised when he lost his job as a teacher in London because he refused to take fees from the poor; he was astonished to be sacked from a Paris art gallery for advising potential customers that they were about to buy rubbish.

And possibly the biggest shock of all came when he was dismissed from his job as a missionary because he had sided with the miners during a strike.

And so this son of a family that did not exactly frown on worldly success found himself without job or prospects at an age when most of his contemporaries were settled householders.

Van Gogh was 27 when he decided that painting might be his causeway out of misery; it is an age at which most artists have left their apprenticeship far behind.

He made a few attempts to learn the craft in the orthodox way — but learned only that most teachers are mediocre painters who dread originality and that most students are slaves to fashion. From both he got humiliation.

It was probably just as well: he might have stayed at school too long — but the belief that he was a self-taught genius who suddenly erupted onto canvas is a myth.

Van Gogh was not a man for half-measures — and when he set about mastering (although he would not have used that word) the craft of painting, he gave himself possibly the most thorough grounding any artist ever knew.

He would not have used the word 'mastering' because it betokens complacency: he used to say art was 'always seeking without absolutely finding' or 'attempting to break down the wall between what you can see and what you can express'.

Far from being a 'primitive' he was acquainted with every theory that had ever been uttered about painting, including those of his own day; he was an assiduous maker of copies from a great variety of sources.

And in another way he had acquired rare capital: as teacher and missionary he had worked with the very poor and knew them far more intimately than did the 'peasant painters' who were fashionable at the time.

He admired Millet but went beyond him: 'The Angelus' is great — 'The Potato Eaters' is less lovely a picture but it is a far more powerful expression of the realities of peasant life.

That was his first masterpiece, the synthesis of the vast amount of sketching and painting he had done in the early years of his new and final vocation.

Then he lived in Holland, continually saddened by the lot of the masses — but aware too that no matter how harsh the world you paint, it takes on a certain beauty and dignity in the depiction.

Out of that period came hundreds of pictures — of people at looms and sweeping streets and carrying sacks of coal and pushing wheel-barrows and loading boats and engaging in all the other activities that underpinned their world.

Almost all these are sombre: the natural darkness is exaggerated by Van Gogh's feeling of oppression. One notable bright painting belongs to that period: it is of trees in Autumn and seems to say that though man is enslaved, they are free.

Eventually Vincent moved to Paris, partly because his brother Theo, on whom he depended for money, lived there, partly because it was, or was supposed to be, the capital of art.

He spent a relatively happy period there and painted some pictures, especially 'Montmartre', that capture the atmosphere of that great city better than any other artist ever did.

It may seem strange that the grace and lightness of Paris should be best expressed by a Dutchman; perhaps it was stranger still that he was to be the great articulator of Provence.

A lot of romantic nonsense has been written about his going to the south: there was no mysticism in his journey; the Winter fogs of Paris were not to his liking — and Toulouse-Lautrec had advised him to go to Arles, if only because its climate made living cheaper.

Nor is it true to say that Van Gogh 'found himself' there — but its dramatic scape of land and sky and its climate corresponded to his own fierce spirit. He found some peace; he worked furiously — and felt that he might be great.

It is not true that he died unrecognised. nor is it true that he never could sell a picture. Those myths, like the cutting off of his ear when in fact he clipped off a little of the lobe, belong to our hunger for melodrama.

He sold pictures — but mostly to friends. There were those who deemed him a genius — but they were guerrilla artists such as himself. The general recognition he yearned for never came.

It is true that in his last few years he was intermittently not in his right mind — the probable cause was the lack of understanding of his work by the world at large.

His confidence ebbed; he began to repeat himself in his work; what had been marvellously original was now only a mode; he knew it — and nothing hurt him more than painting badly. Eventually he shot himself in the chest. It was probably no more than a cry for help — but after a few days he died. He was 37.

It seems the perfect ending for the Bohemian — but Vincent Van Gogh had a horror of that life and yearned for a decent income and a wife and a family. His famous 'Bedroom at Arles' is the cry of the outsider for domestic peace.

It was not to be. His own words, written to Theo, were painfully right: 'My soul is like a house with a few wisps of smoke coming from the chimney — the passer-by does not know what a great fire is blazing within'.

Evening Press, Tuesday, June 7th, 1977

Woody Guthrie - a very strange background, a very rare talent

Woody Guthrie called his autobiography 'Bound For Glory' and was not unknown to refer to himself as a born winner — the strange truth is that he was not being ironic.

But it would be hard for the innocent bystander (if there is ever such a person) to understand why a man so relentlessly tortured should regard himself as a darling of the gods.

Woody Guthrie's life was on the surface a chronicle of hardship and domestic tragedy and finally of long illness and yet it is not an accident that his best-loved songs are 'This Land Is Your Land' and 'So Long, It's Been Good To Know You'.

The truth is that he was a born winner, though not in the sense of fame and fortune; the former came to him late, the latter not at all. Nor was he noticeably successful in his quest for personal happiness.

He was a winner in that he had a magnificent talent and he expressed it with the single-mindedness of Lester Piggott—and like Piggott he may not have been particularly happy but he was to a rare degree fulfilled.

Adler, the least fashionable of the great trio made up of himself and Freud and Jung, would have understood him better than most.

In Adler's summary of psychology he lists three ingredients necessary, though possibly not sufficient, for mental health — satisfying work, a good marriage or its equivalent, and people with whom you can communicate.

Of the three the first is by far the most important — there is a wealth of truth in Thomas Carlyle's: 'happy is he who has found his work'.

Woody Guthrie, himself for a while a kind of rustic psychologist, though possibly of the school of Will Rogers rather than that of Vienna, would have agreed.

Once he had found his vocation he never deviated, even though it meant poverty and humiliation and abandoned marriages and all the misery that having no fixed place on this earth entails.

Saints, according to popular wisdom, are not easy people with whom to live — Woody Guthrie qualified on that count at least.

When he walked out to buy a newspaper or a few cans of beer, his wife might not see him for several months again—even though he intended to be back in a few minutes.

It was not that he was a professional anti-bourgeois: that attitude is a luxury enjoyed only by those who have a good stake in the earth.

He appreciated more than most the need for a house and a bit of land and a regular income—but the possibility of achieving these would have meant a betrayal of himself.

He well knew that most people must remain like trees rooted in the one spot—but he was different, the product of a combination of a very strange background and a very rare talent.

To understand Woody Guthrie's pilgrimage you must know that he was a child of The Dustbowl, of that time and place when modern capitalism was most naked and not particularly ashamed.

He was born in 1912 in Oklahoma and lived there and in West Texas until he finally went walkabout. In his youth that region was ravaged almost beyond repair.

Food and oil were in frenzied demand: the soil was mined rather than husbanded; the oil was extracted with reckless disregard for the local consequences—eventually there was neither earth nor clean water.

The Dust Bowl was a vast area where short-sighted methods of farming had reduced the soil to minute particles; when the winds came, the earth was blown away.

And so one of the most fertile regions of America became a desert and even neighbouring areas that escaped had in Woody Guthrie's words 'a beat-up look'.

Sartre was later to write a famous essay in which he expressed the fascination of the French, the most rooted of all nations, at the way people left home in America and set out on vast journeys into the unknown.

Steinbeck's 'The Grapes of Wrath' is the most notable example — but his protagonists, the Joads, did not really leave their home—what had been their farm had been blown into the Gulf of Mexico.

And so they, as most of their neighbours did, moved westwards to the promised land of California; so did Woody Guthrie — and The Dust Bowl and its attendant miseries influenced him as powerfully as it did Steinbeck.

But Woody moved alone: the family mould had been broken beyond repair. His father, once a prosperous estate dealer, had come shatteringly down in the world; his mother, a beautiful and gifted woman, had after years of intermittent insanity died in a mental hospital; his favourite sister had perished in a terrible accident.

And yet the family legacy was not entirely tragic: the early years had been good—and his father had passed on his courage and his mother her love of music. And the era, crude and savage though it was, had infected him with its energy.

And yet Woody was in his mid-twenties before he set out on his odyssey. He was then a sign-writer in Pampa, a boom town that was dying because all its oil had been extracted.

And so one day he made a bundle of his brushes and started walking and hitching westwards. As pretext he had some vague idea of going to a rich aunt in Sonora; his real impulse was curiosity—he wanted to be part of this strange migrant life.

And he was beginning to understand that he was one of those who are born to write or sing on in some such way express themselves: 'Things were starting to stack up in my head and I just felt that I was going out of my wits if I didn't find some way of saying what I was thinking'.

And so slowly and fumbllingly Woody Guthrie became the bard of the new nation, that composed of an uprooted people, that symbolised by the occupants of an old jalopy in which he gets a lift.

" 'All you folks one family,' I asked them. 'All one family. This is me'n my husband, an' these is all the' kids we got left. Four of us now. Used to be eight.'

'Where's the other four?' I asked her.

'They just went,' I heard the lady say."

And California gave permanent shape to his political intuitions: promised land it was but only for the few who owned it—its great orchards and vineries and gardens were underpinned by the ill-rewarded sweat of the transient labourers.

Woody Guthrie is often called the founder of modern American folk music. The claim is exaggerated: he owed a great deal to the blues and the amalgam of hymns and songs that were the heritage of his people in Oklahoma.

What makes him unique is that he was a great singer and composer who was also an unremitting and utterly pure socialist—so pure that neither Communist party welcomed him.

There have been and are other 'singers of the people' who are far above what Scott Fitzgerald called 'the hot struggles of the poor'—Woody was always down there with them and sometimes a few degrees below.

The man who wrote 'Tom Joad' and 'Union Maid' and 'Dust Bowl Refugee' and all the rest of the thousand odd songs he left behind did so out of an immediacy that was his greatest strength.

He was, of course, eventually 'discovered' and the hot hands of commerce attempted to embrace him—but after a successful audition in a celebrated New York studio, he went to the washroom and did not come back.

The show business whizz kids had been dicusssing how best to package him and when one lady suggested he perform in the midst of happy farm-workers, he decided it was time to return to the real world outside.

He continued to make a living as he had done since hitting out for California and interpreting the theft of his brushes on the road as a sign that his career as a painter was over.

He sang and played in bars and other places where working people might be gathered together; he had learned early that the rich were not too free with their money, even when it comes to throwing a coin into a hat.

Fame, such as it was, brought more respectable employment—but he had a rather disconcerting habit of turning up late or not turning up at all.

It was hardly due to indiscipline: he had served honourably in the war. It was probably an aversion to being lionised, to being lauded for the wrong reasons.

And yet he could not help being influential: the disciples emerged—Pete Seeger and Bob Dylan are probably the best known, though hardly the next of kin.

Few will argue if you call Woody Guthrie the greatest folk singer of all time, though of course 'all time' in this sense really refers to the last 50 years and to the English-speaking world.

He was not the greatest singer, nor is he the greatest song writer—much of his vast output is eminently forgettable. But in breadth of experience and in honesty and understanding he is without rival.

And if he is called the father of modern American folk music, it is because he wrote not only of the fields but of all the places where men and women 'sing in their chains like the sea'.

And what too makes Woody Guthrie remarkable is that he was an affimer: he saw his people suffer the worst obscenities of modern capitalism and yet believed, as did the Joads in 'The Grapes of Wrath', that good would triumph in the end.

It may sound sentimental—as sentimental as his passionate belief in good earth and pure air and clean water, as sentimental as the message in 'This Land Is Your Land'.

What was also remarkable about Woody Guthrie was that he achieved so much in so little time: his major career was over by 1954 when he had to enter hospital, suffering from a progressive disease of the nervous system.

He battled on until his death in 1967.

In a recent interview John Rotten of The Sex Pistols said: 'Music is about fun. You remember fun, dontcha?' Woody Guthrie sure did.

Evening Press, Tuesday, 15th, 1982

James Joyce — spoiled romantic

Let us suppose that James Joyce had never found a publisher and died in obscurity and that someone had inherited his manuscripts and sent 'The Dead' to the New Yorker, would it have been accepted?

Had it been in the reign of Harold Ross, this story that is now generally deemed a masterpiece might have got in — but severely altered.

One can see that dogmatic man wielding the proverbial blue pencil fiercely.

And against the celebrated last paragraph he would probably have put some vulgar and semi-cryptic comment such as "So what?" before going off to winkle Dorethy Parker out of The Algonquin.

It is, of course, an outrageous piece of green prose. What, for instance, are mutinous waves?

But it succeeds — because of its boldness. Joyce is a supreme example of the fact that people tend to take a man at his own valuation.

Some years ago a literary prospector told a rather shocked world that this famous passage is pastiche — and perhaps it is.

And there is no great harm in that. If true, it merely buttresses a suspicion that must dawn on anyone who reads 'Dubliners' without totally suspending the critical faculty: far from being the work of a master, it is a woefully uneven book — and certainly does not show Joyce as the wordsman of popular repute.

There is little need to go into a tedium of examples. One should suffice.

James Duffy as 'A Painful Case' draws to a close is standing on Magazine Hill in The Phoenix Park and sees "a goods train winding out of Kingsbridge Station, like a worm with a fiery head winding through the darkness, obstinately and labouriously".

It is—to say the least of it—an unhappy image and not happily expressed.

And yet 'Dubliners' is a work of genius—if that maltreated word has any meaning left.

Despite the flawed writing, it lives: it abounds in acute perceptions—and through it runs a warm river of feeling for a wide range of people.

It has been said of Joyce as of Guy de Maupassant that he was brilliant but cold.

In both cases the charge can be easily refuted; to acquit Joyce you need hardly go beyond 'Two Gallants.'

Lenehan, the more intimately revealed of its protagonists, is a strange hero — a layabout before the term was minted and something of a lickspittle for bad measure.

And yet you probably remember him more clearly than any other person in the book — and not without affection..

There is nothing externally heroic about the scene in the little back-street restaurant where he invests two pence halfpenny in a plate of peas and a bottle of ginger beer.

But as Lenehan looks into his soul with devastating honesty, you feel for him.

He is thirty, has no profession, and never had a decent job. His Promised Land is a home of his own with "a warm fire to sit by and a good dinner to sit down to."

I suppose you could say that Lenehan is convincingly portrayed because he is an aspect of Joyce himself — just as is Gabriel in 'The Dead.'

But does it matter? That kind of seeking after concordance is a harmless — though for some not unprofitable — pastime; it has little to do with literature.

There is a saying about inquisitive people: "He (or she) would ask you what you had for your breakfast."

Joycean "scholars" (and they aren't all American) have an unholy lust to find out not only what he ate for breakfast, but for every other meal of his life as well.

Again one is compelled to ask: "What does it matter?"

The scholars will answer that it matters a great deal because in the details of his life you will find clues that help to explain how he became the kind of artist he was—and that may unravel some of the obscurities in his writing.

The answer to the first part of that argument is that surface evidence may tell you little about a man's development—the deepest wounds may remain forever hidden.

And the answer to the second part is that a good wine needs no bush—and it's a work of dubious value that needs a key.

But Joyce's adulators are so in awe of him that if he had lived longer and gone further into mystagogery, they would have accompanied him as unquestioningly as the children of Hamelin followed The Pied Piper.

It is perhaps salutary to mention here what one might call the Ern Malley syndrome.

In the thirties in Australia—and elsewhere—there was such a fashion in obscure verse that intelligible poetry was more or less outlawed.

Two students in Sydney decided it was time to cry "Halt" and sent an epic poem to the leading literary magazine of the day—with a note about its origin.

The manuscript, said they, was the work of an adventurous young man named Ern Malley who had died young and tragically away up in the outback.

The poem was published and rapturously received; Ern Malley became a cult—for a while.

The two iconoclasts revealed that they had cobbled the masterpiece together by taking fragments out of a government treatise dealing with the eradication of unprofitable insects — so much for trendy gurus.

There is also the Picasso syndrome — the belief that because a man has proved himself in conventional art, his avant garde work must be good.

Of Joyce's conventional work, his collected verse, "Chamber Music" is best forgotten. And the posthumously published "Stephen Hero" is part of an early version of "A Portrait of the Artist as a Young Man".

The "Portrait" is a brilliant book. It is pretentious—but in a pleasing manner: it is the voice of a young man at a time when he feels his powers are limitless.

It is often praised for its set pieces—and most of all the bitter argument about Parnell that spoils the Christmas dinner—but perhaps its greatest virtue is in its expression of youth's yeast.

One thinks of Thomas Wolfe: "We were young—and we would never die."

And what of 'Ulysses'? Unlike 'Dubliners' and the 'Portrait' it has not been universally acclaimed.

It is true that the dissenting voices are no more than the squeaking of a mouse in a storm—but that they are there at all may shock some people.

The heretics include D. H. Lawrence who protested against the attempt to ban the book—but called it a load of rubbish.

Frank O'Connor and Arthur Power were less explicit—but expressed grave doubt.

And certainly Joyce's claim that 'Ulysses' would capture the essence of Dublin on a given day is preposterous.

It is preposterous not on account of the scale of its ambition — such boldness is admirable — but because for such a task he was palpably ill-equipped.

A knowledge of Dublin's topography and the minutiae of its surface life was not enough.

Joyce's great weakness was that he had little knowledge of the kind of life led by the great majority of Dublin's citizens.

In his short stories and in the 'Portrait' we see his sympathy with a wide range of people — but they are almost all in some degree members of the middle class.

The working class were a closed book to him — and seemingly a book he had little desire to open.

This is not necessarily a flaw in a writer—but it is in the context of the ambition behind 'Ulysses.'

This insensitivity is already obvious in 'Dubliners' — and especially in 'Two Gallants' and 'A Painful Case.'

When Lenehan goes into the little back street restaurant, there are only three customers in it—they are dismissed as 'the mechanic and the two work-girls.'

An incident in 'A Painful Case' is perhaps more revealing.

James Duffy in his lonely wandering enters a public house in Chapelizod where he sees "five or six working men."

It is a decidedly peculiar expression. Five or six? One feels that to Mr. Duffy they were so insignificant that it hardly mattered.

It can be argued that the insensitivity is not that of Joyce but of his characters.

Maybe that is so—but it is hard to avoid the suspicion that he saw the working class as almost a different species.

About his intellectual snobbery there can be no argument.

It is palpable—especially in that much-admired story 'Grace.'

It is a brilliant piece. But it reminds one of what Joyce said about his father—'cruelly shrewd.'

And one is tempted to apply those words to 'Ulysses' too.

It is a brilliant work — but it lacks heart.

It has a greater fault — it fails Desmond McCarthy's famous test.

The great critic's thesis was this: no matter how sordid the life expressed, a work of art should not leave us without a sense of the possibility of a better world.

Edmund Wilson would not agree that 'Ulysses' fails this test.

In his celebrated essay in 'Axel's Castle' he argues that essentially it is a portrayal of man struggling towards a better life.

D. H. Lawrence believed the opposite — that it was obsessed with decay.

Perhaps such questions can never be beyond the subjective—but the fact that we argue about them is a hint that we believe otherwise.

And 'Ulysses' seems to some people to be the work of a failed romantic — it evokes "Life being what it is, one dreams of revenge."

Its stance is alarmingly like that of the public house wit who reduces the world to his own dimensions.

Clever it is — and fascinating, especially for those who have a fair knowledge of Dublin.

But how much of the fascination is extrinsic?

Some would say that the interest of scholars who grew up far from Mulligan's and Sandymount Strand answers that question.

And perhaps it does. And perhaps that scholarly interest is due to the fact that Joyce more than most writers left much to unravel.

Perhaps there is a Thomas Hardy society in Japan — not to mention Milwaukee — but the fields and towns of Dorset are unlikely ever to see such academic exploration as Dublin enjoys.

Hardy was too explicit — he left behind him little thesis fodder.

Nor will you meet many roaming scholars in the pubs of Doneraile or around Twopothouse.

Canon Sheehan was too explicit also — and as an artist he was at best an amateur.

And yet in fragments throughout his novels you will meet powerful intuitions, compelling insights into the totality of Irish life.

Technically 'Ulysses' is enormously bold — but one feels that unlike 'Dubliners' and the 'Portrait' it is not the voice of the full man.

And J. B. Priestley had a point when he condemned 'the stream of consciousness' technique as essentially lazy.

It reminded him of a woman who pads around the house all day in dressing-gown and slippers.

And I am tempted to say that 'Ulysses' is not only a lazy book but a cowardly one.

It seems to me a betrayal of Joyce's better self, the romantic of 'Dubliners' and the 'Portrait.'

Joyce's exile wasn't geographical — like Dylan Thomas he seems to have been trapped in his adolescence.

'Finnegan's Wake' is an entertaining piece of escapism, a kind of 'Loneliness of the Long-Distance Punner.'

You may suspect that these judgements are rooted in prejudice.

They are not: long long ago I risked expulsion from university in his defence.

I believe that the Joyce who wrote 'Dubliners' and 'A Portrait of the Artist as a Young Man' would have done the same in defence of Ibsen.

Evening Press, 1982

Split in the Spanish soul

It is not uncommon for people to spend a few weeks in Ireland — and write a book about us.

And especially if the writer is critical, we tend to accuse him of arrogance and pretension. How could he say so much after so short a stay?

And yet it is not as simple as that.

There are people who have been born in this country and lived for long years here — and yet are criminally ignorant of its history and its potential.

And I know no more perceptive book about Ireland than Evelyn Hardy's "Summer in Another World" — she wrote it after only a few months in West Cork.

The open-minded visitor sees things that we take for granted — it is like the eye of childhood.

And when you go to a foreign country, you are in a sense re-born.

The skin of habit is peeled from your senses—you are as expectant as a child going to his first circus.

Spain for me was a marvellous experience — and both gave a glimpse into Ireland's future and provided a kind of living museum of its past.

The future I glimpsed may prove illusory—but when you see how people can live in the daunting surroundings of Castile, you cannot but marvel at Ireland's natural riches and hope that some day our talents will be taken from under the bushel.

The museum of Ireland's past is in the Spanish people — they are gentle and gracious in a way that for me recalls the generation of my childhood.

This may seem a contradiction—because history will record the thirties in this country as a turbulent era. And the Spanish Civil War was even more savage than ours.

But the contradiction is only superficial — between man the social being and man the political being there can be a gulf that hints at the complexity of human nature.

195

And certain aspects of behaviour in the Spain of today must for an Irish visitor of my generation seem like products of the time machine.

I am talking about such sights as families walking together in the evening.

You will see this not only in the towns — you will notice it even more in the countryside.

One of the great pleasures of the peasant's life is the contemplation of work well done — it provides his little furloughs.

And in the evening hours you can see whole Spanish families—from grandparents down to toddlers — walking around their gardens and olive groves and fields.

And there is a marvellous sense of intimacy about these walkings — its outward expression is in the linking of arms and the holding of the hands.

You sense too the pride they take in their land — poor though so much of it is.

In the two smaller towns I briefly knew — San Rafael and Segovia — there is a similar love of walking and a similar sense of intimacy.

And the plaza is a most civilised institution.

The plaza in Segovia is especially beautiful: it is bordered by trees and cafés — and the citizens sitting at its tables look as if they feel that they are at the centre of the world.

We all occasionally experience something that—slight though it may seem—we feel will lodge forever in our memory.

And one evening in the plaza in Segovia I had such an intuition.

A woman who had grown old gracefully—she was tall and erect and white-haired and handsome and would have made a marvellous Moore Street queen—was making a speech to her seated peers.

I wasn't near enough to guess what she was saying—but obviously it was funny and probably ribald.

She sat down to great acclaim—not all of it reverent. And I couldn't help feeling that the plaza is a great antidote for loneliness.

The plaza, of course, is an institution of warm climates — we can only envy its beneficiaries.

The Spanish have another advantage over us—the motor car hasn't yet torn their social fabric.

Again you are taken back to the Ireland a generation ago.

And you notice the change most keenly in the decline of 'the field', an institution still flourishing in Spain.

'The field' needs no explanation to lovers of 'Knocknagow' — it was the rural focus where in Summer evenings the males of the tribe disported themselves, sporadically watched by the females.

In Ireland 'the field' is now little more than a memory — in every few miles of the road in Spain you can see the goalposts and the local Mat the Thresher and his comrades.

Most of the goals, incidentally, have nets—and for a good reason: they are to prevent the ball from going over a cliff or into a river or otherwise being lost.

And what about the ball that goes wide? The local joke is that shots are essayed only from very close range—and that this topographical necessity is reflected in the play of the national team. But that's another story.

Another obvious agent of social bonding is the rural pub. Again the comparative paucity of cars is a factor—you will find a busy bar with few cars outside.

And the Spanish pub—or café as it is more correctly called—is as different from its Irish counterpart as a garden is from a yard.

For a start the beer is far better—ours in general is a tribute to the powers of a near-monopoly.

The more obvious difference is that the café is more than a drinking place — the humblest carry a fair array of food.

It may not be exotic food — but hard-boiled eggs and omelettes and ham and a variety of fish can keep body and soul together if only for a little while.

In the typical Irish rural pub you will do well to get a ham sandwich, usually one that has seen better days.

Another thing I liked about these Spanish pubs was their architecture: they are simple in design—and built from local wood and stone.

And, heaven knows, there is no scarcity of wood and stone in the country between Segovia and Madrid.

There are few tracts of that brown semi-barren land unpeopled by trees; they have two functions — they temper the force of the sun and they provide wood for fuel and building.

And you are never far from rock — in most of central Spain the jagged mountains dominate the landscape.

Irish mountains are generally soft of contour, almost benign; their Spanish cousins are uncompromising.

Once upon a time I used to think that Lorca's poetry was overblown, pretentious, at times melodramatic.

Now I feel I understand it better: he was a child of Granada—and that is mountain country too.

And when you have passed through that forbidding landscape in Mid-summer heat, you know why the people worship water.

In Madrid too you see the brutal extremes of wealth and poverty that generated the Civil War.

197

On the outskirts of the city is The Escorial, Spain's greatest showpiece for tourists and a memorial to the heyday of the Empire.

This one-time royal palace is a work of stunning beauty.

Not far from the fashionable heart of Madrid are hovels that chill the soul.

And you cannot forget the Civil War—and the proud record of the Madrilenos.

They resisted the Fascists as fiercely as their forebears had resisted the forces of Napoleon.

And the capture of the city was looked on by Franco as the key to the war—and not only for strategic reasons.

The most celebrated engagement of the war was fought to keep open Madrid's last life-line—the road to Aragon and Catalonia.

This was the battle of Jarama — that resonant name is one of the most evocative in socialist folklore.

No poem or song can exaggerate the heroism of that battle — the British Battalion, for instance, spent three months in the trenches without relief.

Most of them were only amateur soldiers—but they held out against Franco's professionals and held up the capture of Madrid for two years.

Among those who died there was the legendary Dublin republican, Kit Conway.

And those who expected the Madrilenos to be hostile to the British visitors to the World Cup didn't know their history.

There was the usual friction among the young — the older Spanish knew better.

The common people haven't forgotten the help that came to them in their years of trial.

But it would be sentimental to look on the Civil War as if all the good and the gallant had been on the one side.

War is seldom like that. And the socialists lost eventually not only because most of them were amateurs but because there was a certain dishonesty at the core of their movement.

If they had won, Spain's freedom would have proved illusory: it would have led to a communist dictatorship — and probably another civil war.

George Orwell saw that—and hence had great difficulty in getting a publisher for 'Homage to Catalonia'.

Picasso's 'Guernica' stands as a record of Fascist obscenity — but terrible deeds were done by the other side too.

The regular forces of the socialists fought chivalrously — but the peasants' revolt was savage.

Centuries of repression bred a bitterness that led to terrible and sometimes arbitrary vengeance.

The country between Madrid and Segovia saw the fiercest of the Civil War.

There lies the Jarama Valley — and the town of San Rafael, another resonant name in Socialist folklore.

And some of the grandfathers I saw tenderly holding the hands of little boys and girls know all about the foul rag-and-bone shop of the heart.

And if you ask a Spanish philosopher why Franco's regime endured so long, he might answer that the Civil War had left a horror of violence.

A similar reason is sometimes given for the passivity of the Irish during The Great Famine—the memory of 1798 was too recent.

Francisco Goya was ahead of his times not only in technique but in his insight—he expressed the violence in the Spanish soul.

And his own life was riddled with the contradictions that Spanish philosophers are forever examining in their fellow countrymen.

He was a court painter who stripped the rags from Spain's sores.

Goya was a son of the mountains too—he was from Aragon and that rugged province left its mark on him.

And he lived in a Spain that seemed to have lost its way—so much so that he probably welcomed the French invasion.

This tortured man spent his last years in Bordeaux—and in France he seemed to find the sense of wholeness that had so long evaded him.

The second-last picture he ever painted, the famous 'Milkmaid', is sad—but in a tranquil way.

Allow me one last observation on Spain.

I had often heard that the Spanish are lazy—maybe they are but, to mint a cliche, you could have fooled me.

In the hotel where I stayed in Segovia the staff who tended bar until well past midnight were up and around at breakfast—and worked all day.

And in Madrid in noon heat of 100° I saw men carrying the hod and digging trenches—and they had no siesta.

And so forgive me if I suspect that the popular concept of Manana is about as authentic as that of the leprechaun.

Evening Press, Tuesday, May 31st, 1983

Voice of a hidden world

In Britain—and especially in England — the term 'common people' is not used lightly.

Nor is 'lords and commons'—it stands for an old and deep distinction.

Most histories of Britain tend to stress its role in the world context—and to neglect the domestic struggles.

And thus a lad or lass can go out into the world rejoicing in Trafalgar and Waterloo and The Charge of the Light Brigade—but knowing little about the upper classes and the common people.

They may have come across a few paragraphs about Jack Cade and Wat Tyler and Jack Straw — but are unlikely to have read about Feargus O'Connor and Joseph Arch and The Tolpuddle Martyrs.

It cannot be said too often that the British ruling class treated the common people of their own country as badly as those in the lands they annexed.

Indeed, they treated them worse than they treated ours: Ireland escaped the Industrial Revolution—and the workers in the fields had the furlough of the night.

Except for the last quarter of the nineteenth century the British rural workers were never effectively united — their Irish counterparts were never united at all.

It wasn't until the 18th century that the British labour movement took discernible shape: until then it had produced no more than sporadic and pathetic revolts.

It began mainly in the mines — and its Johnny Appleseeds were the preachers unleashed by John Wesley.

It would be hard to overstress their role in the making of modern Britain.

'The non-conformist conscience' is another term not to be taken lightly.

D. H. Lawrence grew up in mining country—it was dominated by 'the chapel'.

In an essay on the preachers he calls them 'cheeky' — 'bold' might be a better word.

They were perhaps narrow—but they had the strength that often goes with narrowness.

They were not inhibited by the suspicion that somewhere lurked spiritual and intellectual fathers: they were their own men.

And above all they were republicans: they detested privilege. The survival of the Welsh language is a tribute to their sense of values.

And their republican outlook presents a seeming paradox: the monarchy could hardly survive without their support.

Some people may look on the British—and especially the English — as lovers of royalty.

That is too simple a view: it isn't long since they not only dethroned a king but executed him.

Charles the First wasn't a bad man—and he was certainly a brave one; however, he underestimated the power of the emerging middle class.

200

And so on a snowy January morning in 1649 he found himself on a scaffold in Whitehall.

The monarchy almost fell again early in this century: Edward, the royal layabout, was despised by the common people—but he had the good sense to die before they dethroned him.

They wouldn't have marched on London like Jack Cade and Wat Tyler and Jack Straw — 'public opinion' is another term not to be treated lightly in the British context.

The monarchy survived — and three of those who occupied the throne since the royal layabout have comported themselves with dignity—but that is only part of the story.

One of the reasons for its survival is obvious: people like to believe that somewhere there is a life better than their own.

And that is a joke that must cause Elizabeth to perpetrate the odd wry smile: the poor woman is more or less a prisoner; she will never go into The Kings and Keys for a pint of ale — or even a half pint.

There is a less obvious reason for the survival of the monarchy.

The common people look to it as the guardian of their rights.

And high among these is the right to common land and the right of way.

In this they have much to teach us: we are great people for speaking about abstract freedom — but not so good at safeguarding it in the reality.

Our entrepreneurs are experts at nibbling at common ground.

And many paths that by usage had become more or less rights of way have disappeared.

The ubiquity of the car is one cause: short-cuts to the bog or from one road to another are almost forgotten.

And fishing paths are endangered by the decline in the standing army of Irish anglers.

Our neighbours are fierce watchdogs of these rights.

Tomorrow they will picnic on Epsom Downs with a sense of ownership as total as if in their own gardens.

And woe betide the farmer that interferes with a right of way: he will know the kind of wrath that caused Edward the layabout's throne to totter.

Thomas Aquinas believed that farmers shouldn't own the land but keep it in stewardship for the common good; that ideal is now sadly tattered — but the common people of Britain are determined to keep it alive.

And they have much to safeguard: you can walk the length and breadth of he island — and seldom touch a road.

The start of this century brought the great walking years.

A century is an arbitrary division of time — but people do not see it that way.

And it was natural that the year of Our Lord and Queen Victoria 1900 seemed a threshold.

It produced people who looked on themselves as a new breed — and in the main it was a revolution of the lower middle class.

H. G. Wells, son of a professional cricketer, was its chief apostle — his novel, 'Ann Veronica', could be said to popularise ideas half-buried in 'Jude the Obscure.'

The revolution could be interpreted as more than a response to the new century.

It was a reaction against the Victorian values — or at least the popular concept of them: Victoria herself was hardly 'Victorian' — and may, indeed, have inspired 'Lady Chatterley's Lover'.

And it was, of course, a reaction against The Industrial Revolution, against the dark Satanic mills.

William Wordsworth's doctrine was reiterated — and the bridle-path and the right-of-way became the roads to the new Samarkand.

And the symbol of the quiet revolution was the walker, with a volume of poetry and a note-book and a sketching pad in his knapsack — and belief in a better world in his soul.

He explored the world beloved of Gilbert White and Richard Jefferies, lunched by hedge or tree on bread and cheese and a bottle of ale — and took his supper and his ease in a congenial inn.

It was a good life — if you could afford it. Some could — more or less: the occasional guinea for an article or a review could be stretched a long way.

Edward Thomas hoped to live thus — and did for a while — but he married young: for the rest of his short life the publisher's dead-line hung over him like Damocles' sword.

Thomas was the son of an earnest Welsh civil servant. He grew up in Battersea, a township that is like a living museum of Englishness and middle-class Englishness especially.

Its values made little appeal to Thomas: his life was to be a search for he knew not what; he sensed there was a key to his destiny — he was never to be sure whether he had found it.

Sometimes he thought of himself as an uprooted Welshman — but this intuition probably over-stressed the power of heredity. Ironically, he is now deemed the most English of poets.

His perhaps mystic notion of his Welshness received little encouragement from his father, a man who had drunk deep from the Victorian chalice.

Edward grievously disappointed him — and the familiar father-son battle left its scars on both. The younger man suffered the more — because so unsure.

He went his way — and 'settled down' (never were words so wildly inappropriate) to a life of writing essays and reviewing books.

The pickings proved too thin to support a wife and children — and he was forced to write commissioned works.

And so he experienced the new Grub Street: it was as if an artist had to make his living by painting walls.

Perhaps he gained something — if only discipline: Theodore Dreiser, creator of 'An American Tragedy', said he learned to write in a peculiarly American version of Grub Street.

A Chicago publisher employed him to complete manuscript novels that had been cut in two: he added a new first half to one part — and a new second half to the other.

Thomas at least learned to write quickly — and not all the commissioned books lack distinction.

His anodyne in those years of captivity were his essays: they were written for no one but himself.

That is probably an exaggeration: almost every writer is conscious of an audience — even if it is made up of only a few people he knows to be his spiritual kin.

Thomas was a writer's writer, like those footballers who never became famous but are greatly valued by their peers.

Thus he came to the notice of Robert Frost: They met — and the consequences were remarkable.

The supposed sage from New England had a genius for wrecking people's lives — but he brought Thomas nothing but good.

He told him that his essays were the stuff of poetry — they needed only to be formalised.

Thomas dubiously took his advice and wrote his first poem — or rather gave a metrical arrangement to an essay-story he had composed.

And so in his mid-thirties and after fifteen years as a writer he discovered that he was a poet.

This story has been told so often that people tend to doubt it — but it is true.

In the few years that were left to him he joined the immortals — and yet it is almost certain that he looked on himself as a failed man.

He experienced little fame — but obscurity hardly worried him. Indeed it probably made him feel kin with those he admired most.

They were the little people — the farm labourers and charcoal-burners and carters and dairymaids and all the rest that underpinned his world.

His longest poem — and it isn't very long — its a wry tribute to an imaginary hero who in his life had known the totality of the rural experience.

Of course it included going to war: the British ruling class have always been extremely generous in giving the common man the chance of a glorious death.

Thomas's admiration for the common people may seem a trifle sentimental — a little story to which he was a witness suggests otherwise.

When the First World War began, he was living in Hampshire, mainly because he was able to get his children into a school there inexpensively.

The school was Bedales, famous for its pioneering in co-education and other allegedly progressive ideas.

It was run by alarmingly high-minded people — and it is not unfair to say that they looked on themselves as above the common herd.

They did not approve of war — nor did he. However, he didn't relish the prospect of a German victory — and lost no time in enlisting.

The staff of Bedales — and it was a big staff — all embraced pacifism.

However, they were ready to do their bit.

And so one night they summoned the women of the village to a meeting — and assured them that they would dig and cultivate their allotments while their menfolk were away at war.

The women of the village listened. It must have been a mutely hilarious occasion.

You could hardly blame them for suspecting that the good people of Bedales were promising more than they would — or could — deliver.

And so it came to pass. The women of the village dug their allotments — the men and women of Bedales went their high-minded way.

The potatoes and the cabbage and the carrots and the parsnips and the peas and the beans were put into the ground not to mention the lettuce and the cress and the herbs and the onions and the vegetable marrows.

And they grew, and the war was won.

For the Dubliners on their 25th Birthday, 1987

I grew up in a cultural desert. It was the era after the Civil War; bitterness and disillusion stalked the land.

Spiritual values went underground. The Irish people became obsessed with survival. A barren materialism prevailed.

It was accentuated by the economic climate that followed the Wall Street crash-this disaster affected even the Irish peasant farmers.

And a new and more terrible form of puritanism gripped the country; it was the puritan vices without the puritan virtues.

Anything that wasn't seen to contribute to material survival was deeply distrusted.

Music became suspect. Books were deemed likely to lead the readers astray.

Fiddles and melodeons and concertinas and flutes and tin whistles remained silent. Books were hidden in stacks of turf and in ricks of hay and in trees that the elders of the tribe could'nt climb.

New folk sayings sprouted such as "Nothing ever good came out of music"and "Novels are the work of the devil".

And some of our spiritual guardians chipped in and brought an end to the crossroads Ceilidh and the house-dances-both had long been a part of Irish rural life.

And of course this vacuum was filled. Sentimental ditties replaced the old songs. "Did Your Mother Come From Ireland?" ousted "The Rocks of Bawn".

To a struggling people the new songs-such as the were-represented a better world.

They associated the old music with hardship and backwardness. This attitude also hastened the decline of the Gaelic language.

And thus too the bog oak cupboards and the sugawn chairs and the deal tables were banished-they were remainders of the dark ages.

Even the humble mug from which generations had imbibed fell victim to the madness; it was deemed uncouth-the cup took over.

Even as a small boy, I sensed that there was something gravely wrong.

I felt that there was a dimension missing from our world; a sociologist would say that I was suffering from cultural deprivation-but I didn't know the term then.

All I knew was that I needed something of great importance – I was like an otter in a land where all the streams had gone underground.

I knew that the songs of the day, good though some of them were, did not express our world- I was seeking mirrors.

And then one day at a fair in my home town I found a mirror- or at least a fragment of a mirror.

A travelling musician named Batt Coffey was playing a battered accordion and singing "The Rose Of Mooncoin" – for me it was a revelation.

I followed him around the fair until I had memorised the words, even though I had only a few coppers to drop into his little daughter's hat.

"The Rose Of Mooncoin" is hardly one of the world's great folk-songs but it is fresh and honest- to me it was like eating wholesome bread after a surfeit of confectionery.

I wouldn't dream of attempting to say what folk music is-but, like love, you recognise it when it when it smites you.

In a sense there is no folk music-because music is as universal as the air we breathe.

Sean-nos singing is deemed to be the oldest and the purest of all Gaelic music but I have heard its counterparts in a shepherd's hut in Sicilly and in an Indian Café in Shepherds Bush.

One of The Dubliners' great virtues is their freedom from dogma-they do not theorise.

And if they have an unexpressed philosophy, I suspect that it is akin to the ambition formulated by Ernest Hemmingway- "To write nothing that might go bad afterwards".

And in a sense The Dubliners have no special style. Their style is the absence of style-every number is treated according to its nature.

Their great strength lies in their honesty-there is no pretension, no affectation. They may have walked with kings-certainly they have never lost the common touch.

And whence came The Dubliners? Did they spring fully-fledged out of O'Donoghue's pub in Dublin's Merrion Row?

Far from it-their struggle towards the light was like an Irish Spring, marked by false beginnings.

They were, however, lucky in their times; Their apprenticeship coincided with the great revival of folk music.

For generations the flames had been tended by devoted spirits such as Vachel Lindsay and Joe Hill and Woody Guthrie.

They were like Johnny Appleseed, the legendary pioneer who travelled all over the United States, putting down the origins of his beloved fruit.

And there were the collectors who saved so much of what might have been lost, such people as Pete and Peggy Seeger, musicians and musicologists.

In Ireland we had Seamus Ennis and Sean Mac Reamoinn and Aindreas O Gallchoir and Ciaran Mac Mathuna.

But for long years such diligent seekers found 'fit audience but few'.

Ironically, Seamus Ennis was doing a great series for the BBC in the fifties when some bureaucrat decided that it had no future and terminated it.

Within a year the little streams had coalesced into a flood; the great revival was on the way.

The catalyst was probably The Civil Rights movement in The United States; it sprouted songs and attracted some fine singers.

In Ireland the Clancy Brothers and Tommy Makem were great spreaders of the new gospel - they helped to make folk music fashionable.

And about the same time television came to Ireland-1962 will eventually be seen as a watershed in our history.

Ronnie and Ciaran and Luke and Barney were then very young-the tide favoured them.

Within a few years they had gained the respect and affection that previously had been confined to stars of the sporting world.

And when John joined the team as anchor man, the upward curve of the graph became steady and sure.

In the meantime there has been sadness and tragedy; Ciaran is now on the shore, glimpsing his ship at sea; Luke is with The Dubliners in spirit.

The group have been fortunate in getting such splendid replacements as Jim McCann and Eamonn Campbell and, on a permanent basis, Sean Cannon.

The show goes on and on and on-it is hard to believe that The Dubliners have now been with us for a quarter of a century.

Stamina is an element of professionalism-but so is freshness; like Alice in Wonderland you must run fast if you wish to remain in the one place.

The Dubliners have achieved this rare coalition of stamina and freshness; Their appeal is still as great as it was a generation ago when Irish television and all the world seemed young. And even though folk music is as universal as the air, it also tends to reflect its surroundings-rather as fish take their colour from the water in which they swim.

Musically The Dubliners may have no particular style but their attitude to life is distinctly Irish.

It could be described as decently ribald and toughly tender-and, above all, it is humorous. And they can be serious without being solemn.

There is many a flood gone under the bridge and many a pint gone down the red lane since the fair day in my home town when I listened to Batt Coffey-but my heart is still as receptive.

And I think especially of a morning a few years ago when fate cast me off the Holyhead train at about four in the morning in Euston Station.

Earth can hardly have a sight more calculated to daunt the spirit.

All around were the human flotsam sleeping on the floor, wrapped in newspapers. Those who had not abandoned the fight were drinking exceedingly weak tea out of paper cups-you pitied the waitresses who had to serve in this waste land.

That morning I felt that the world was meaningless and that my life would never again pick up-until I heard a man singing.

He was a little brown Indian, once no doubt a peasant but now condemned to tidy up the tea-bar in Euston Station.

And as he moved around amidst the debris, he crooned softly to himself; I did'nt know a word of his song-but it was beautiful.

I have no doubt that it was about his own lost world, the skies and the fields and the trees and the rivers that he might never see again. It was his anodyne - it became mine too.

And whenever I feel that life is meaningless, I think of that little brown man.

The Dubliners too would have loved him.

Long may they speak for us.

————

Programme Note for Luke Kelly's Memorial Concert

The Song of A Bird Alone

Giusseppe Ungaretti, the great Italian poet who died a few years ago, lived for a while in Paris-there one of his friends was an exiled Arab who took his own life.

He wrote a little poem in his memory. It ends: E non sapeva sciogliere il canto del suo abbandono- he wasn't able to express the song of his loneliness.

Luke Kelly lived much of his life among his own-and yet he too knew a kind of exile; it was the spiritual exile of one who wished for a better world. Luke was luckier than Ungaretti's friend: he found expression for his loneliness.

For him singing was as essential as it was for the American blues singers who found themselves 'lonely and afraid in a world they never made'.

Luke brought home to you that singing had been man's primal mode of expression. When language was rudimentary, the musical notes expanded it: man sang before he spoke.

And it is fair to say that Luke was a primitive, in the sense of the term as it is applied to such painters as the Douanier Rousseau.

The Douanier was a sophisticate, intimately acquainted with the history of art and acutely aware of contemporary movements- and yet he seemed to have come out of nowhere.

Better than most, Luke Kelly knew what was happening in his own field-and indeed in the fields around it-but he was his own man. Ewan McColl may have inspired him-but only to go his own way.

It may seem paradoxical that such a 'bird alone' should gain fame with a group-but the Dubliners were less a group than a meitheal. In the old peasant pattern the meitheal came together to do a job- and that was it.

The Dubliners were all individualists- Luke and Ronnie and Ciaran and John and Barney were leaves from different trees, blown together by the wind that changed the world of music a generation ago.

What they had most in common was artistic honesty. Luke's ambition was to express 'the song of his loneliness'. He succeeded as much as a mortal can-and in doing so he became an immortal.

———

MORE THAN A GAME

SELECTED SPORTING ESSAYS

CON HOULIHAN

Definitive collection of sports writing – on subjects ranging from Gaelic games, soccer and rugby to hare coursing and cricket – by the acknowledged master of the craft.

'The writing is impeccable . . . The pen portraits are little gems . . . Proves that sports journalism, when done well, can be a real art form'

Books Ireland

'A wonderful read, and not just for the sports nuts'

Ireland's Own

Paperback: €14.99 | ISBN: 978–0–9545335–0–2
Hardback: €25 | ISBN: 978–0–9545335–1–9
Available from all good bookshops and from www.LibertiesPress.com